'The militarization of space still needs to be better unders
economic, and legal dimensions. This book provides new essential and innovative responses.'

Prof. Kai-Uwe Schrogl, *President,*
International Institute of Space Law (IISL)

'In the shadow of great power competition in space, pan-European perspectives
provide a middle-of-the-road alternative. In this deeply provocative and
thoughtful text, the authors confront the need for a unified European space
defence policy in an increasingly defined space warfighting domain while
preserving as much of the peaceful, shared responsibility of a space-dependent
world as is plausible.'

Dr. Everett Dolman, *US Air Force's Air War College (AWC) and*
the US Space Force's West Space Seminar (WSS)

The Militarization of European Space Policy

This book is focused on militarization as the nucleus of EU space policy and the interrelatedness of European security, industrial competitiveness, and military capabilities in the shaping of this policy.

The EU and key member states have increasingly joined the US, China and Russia, among others, in regarding space assets as critical military, as well as economic, industrial, and technological, enablers. This book tackles this issue by, first, shedding light on the military aspects of EU space policy, with special emphasis on the security and defence dimensions of projects such as Galileo, Copernicus, Space Situational Awareness, and Satellite Communication. In this context, contributors confront the empirical aspect of developments, including the role of different institutional actors and the involvement of specific member states. Further, the volume analyses the discursive, ideological, normative, and theoretical foundations of the use of space by the EU for strategic purposes, drawing on the broad spectrum of European integration/International Relations theory. Last, but not least, the volume discusses initiatives outside the EU by key global space players, with an emphasis on the US and transatlantic space relations. All chapters maintain a solid empirical foundation, in the form of geographical or issue-related focus, with an area-specific emphasis on the EU as a whole, transatlantic relations, the policies of key member states (such as France and Italy), and core space powers such as the US, China and India.

This book will be of much interest to students of space power, security studies, European politics and International Relations.

Thomas Hoerber is Professor and Jean Monnet Chair in European Studies, Director of the EU*Asia Institute at ESSCA School of Management, Angers, France. He is the author/editor of fifteen books.

Iraklis Oikonomou is Independent Researcher based in Athens, Greece. He holds a PhD in International Politics from the University of Wales, Aberystwyth, UK. He is the editor of four books.

Space Power and Politics
Series Editors: Thomas Hoerber, ESSCA, France and Mariel
Borowitz, Georgia Institute of Technology, USA

The Space Power and Politics series will provide a forum where space policy and historical issues can be explored and examined in-depth. The series will produce works that examine civil, commercial, and military uses of space and their implications for international politics, strategy, and political economy. This will include works on government and private space programs, technological developments, conflict and cooperation, security issues, and history.

A European Space Policy
Past Consolidation, Present Challenges and Future Perspectives
Edited by Thomas Hoerber and Sarah Lieberman

Security and Stability in the New Space Age
Alternatives to Arming
Brad Townsend

European Integration and Space Policy
A Growing Security Discourse
Edited by Thomas Hoerber and Antonella Forganni

The Commercialisation of Space
Politics, Economics and Ethics
*Edited by Sarah Lieberman, Harald Köpping Athanasopoulos
and Thomas Hoerber*

The Geopolitics of Space Colonization
Future Power Relations in the Inner Solar System
Bohumil Doboš

The Militarization of European Space Policy

Edited by
Thomas Hoerber &
Iraklis Oikonomou

Routledge
Taylor & Francis Group

LONDON AND NEW YORK

First published 2024
by Routledge
4 Park Square, Milton Park, Abingdon, Oxon OX14 4RN

and by Routledge
605 Third Avenue, New York, NY 10158

Routledge is an imprint of the Taylor & Francis Group, an informa business

British Library Cataloguing-in-Publication Data
A catalogue record for this book is available from the British Library

ISBN: 978-1-032-13744-5 (hbk)
ISBN: 978-1-032-13745-2 (pbk)
ISBN: 978-1-003-23067-0 (ebk)

DOI: 10.4324/9781003230670

Typeset in Times New Roman
by codeMantra

The Duty to Protect

We have the duty to protect life.
We have the duty to protect ordinary lives that can be so
great in their essence.
We have to give the means to protect the institutions that
protect us.

These means must never be used for anything else but to
protect.
And we have to make sure that that is inscribed in the
institutions that hold that power to protect.

It must be so clearly said that all temptation and
corruption will be revealed as petty forces that soil this
sacred duty to protect.

Thomas Hoerber, 26.2.2023, Angers

Contents

Contributors

Mariel Borowitz is an Associate Professor in the Sam Nunn School of International Affairs at the Georgia Institute of Technology and head of the Nunn School Program on International Affairs, Science, and Technology. Her research deals with international space policy issues, focusing particularly on global developments related to remote sensing satellites and challenges to space security and sustainability. Her book, "Open Space: The Global Effort for Open Access to Environmental Satellite Data," published by MIT Press, examines trends in the development of data-sharing policies governing Earth observing satellites as well as interactions with the growing commercial remote sensing sector. Her work has been published in *Science*, *Strategic Studies Quarterly*, *Data & Policy*, *Space Policy*, *Astropolitics*, and *New Space*. Her research has been supported by grants from the National Science Foundation, the National Aeronautics and Space Administration, and the U.S. Department of Defense. Dr. Borowitz completed a detail as a policy analyst for the Science Mission Directorate at NASA Headquarters in Washington, DC from 2016 to 2018. In 2022, she testified to the U.S. House of Representatives Subcommittee on Space and Aeronautics in a hearing titled, "Space Situational Awareness: Guiding the Transition to a Civil Capability." Dr. Borowitz earned a PhD in Public Policy at the University of Maryland and a Masters degree in International Science and Technology Policy from the George Washington University. She received a Bachelor of Science degree in Aerospace Engineering from the Massachusetts Institute of Technology.

Antonio Calcara is FWO researcher at the University of Antwerp. His research has appeared in *International Security, Security Studies, Review of International Political Economy, Governance, Journal of European Integration, and European Security*. He is the author of "European Defence Decision-Making: Dilemmas of Collaborative Arms Procurement" (Routledge).

Thomas Hoerber is a Professor for European Studies and holds a Jean Monnet Chair in European Studies. He directs the EU-Asia Institute at ESSCA School of Management.

Dr. Helen Kavvadia is a researcher in residence at the University of Luxembourg. She was a senior adviser at the European Investment Bank. Her research and publications focus on political economy and economic diplomacy. She has a special interest in space issues and regional development banks.

Major General (Ret) Pascal Legai, started his career as a mission preparation officer in the French Air Force. He acquired experience mainly in the fields of geography, imagery, international relations, Space and Security issues. He has extensive international experience in the geospatial information domain. He had been appointed as a geographer in the French Air Force staff in Paris for 5 years (1994–1999). He was also the Head of the French Imagery Intelligence Centre (2004–2006), Human Resources adviser of the French Air Force (2006–2008), and the Commanding Officer of the French Air Force Base in Grenoble (2008–2010) before joining the EU Satellite Centre in July 2010 as Deputy Director. He has been elected by the Member States as the EU Satellite Centre Director on the 1 January 2015, an imagery analysis centre, based near Madrid, Spain, providing Geospatial Intelligence (GEOINT) and Imagery Intelligence (IMINT) products and services to support the EU external action. He left this position on 30 April 2019. On 1 May 2019, he has been appointed to the European Space Agency (ESA), first as a Senior Security Adviser to the ESA/Earth Observation Director, in Frascati (Rome), and today as a Senior Security Coordinator to ESA DG in Paris. He holds a PhD in International Relations, a law degree, several master degrees in Imagery Processing, Computer Sciences, in History, in British Civilization, in Mathematics. He is an aerospace engineer and also holds an Engineer Diploma in the field of the geographic sciences.

Iraklis Oikonomou is an independent researcher based in Greece. He holds a PhD in International Politics from the University of Wales Aberystwyth.

Lorna Ryan, PhD, is an academic at the School of Policy and Global Affairs, City, University of London. Her interest in space policy, particularly EU space policy, and its evolution, developed from a wider research interest in the governance of EU research. Current research activities include the role of different EU institutions in the development of EU space policy, including its regional dimension. She is a Research Fellow at the EU*Asia Institute, ESSCA, France.

Frank Slijper (1970, MA in Economics, University of Groningen) has investigated military-industrial issues ever since writing his master thesis in 1993. With a focus on arms control, he has also published extensively on the militarization of the European Union, including the role of the arms industry, the European Peace Facility, the European Defence Agency and the EU's military space policy. More recently, the emergence of increasingly autonomous weapons has been central to his research and advocacy at Dutch peace organisation PAX, where he has worked since 2014.

Isabelle Sourbès-Verger is a Research Director at the CNRS (National Center for Scientific Research) in Paris. She is one of the few French researchers specialised in the study of outer space activities and related national space policies. Her research focuses on the interface of national public policy and technological ambitions taking into account very different issues such as strategic matters and international security, the impact of public opinion and the role of media as support to space activities. Beside Europe, Russia, Japan, China and India are the major space powers on which she developed extensive expertise. Using comparative analysis of different national space policies, her research provides an original view on various models of acquisition and development of national space competences. Her current work deals with the part to be played by space capabilities, including space surveillance, in the construction of European security and defense architecture. She is a member of different international networks such as European Space Policy Institute Network (ESPI), Space Security Working Group (McGill University). She contributes to different advisory panels at the national and international level. Currently she is part of the Organisation Committee of Manfred Lachs international conference on "Global Space Governance", Montréal, 2014, 2016 and 2017. She is also a member of the ESA History Project Academic Council (EHPAC). She has authored or co-authored in French and English books including The Cambridge Encyclopedia of Space, more than 60 chapters or articles, about 20 research reports for French and European decision-making bodies.

Dr Dimitrios Stroikos is LSE Fellow in the Department of International Relations at the LSE and Head of the Space Policy Programme at LSE IDEAS. He is also the editor-in-chief of *Space Policy: An International Journal*. His research publications have appeared in journals such as *International Politics, Review of International Studies*, and *Journal of Contemporary China*. His latest publication is "International Relations and Outer Space", *Oxford Research Encyclopedia of International Studies*.

Dr. Jessica West is a senior researcher at the Canadian peace research institute Project Ploughshares. Her research and policy work focuses on technology, security, and governance with a particular interest in peace and security in outer space. Jessica interacts regularly with key United Nations bodies tasked with space security and space safety issues. She holds a PhD in global governance and international security from the Balsillie School of International Affairs, Wilfrid Laurier University.

Introduction

Thomas Hoerber and Iraklis Oikonomou

In their last book, Hoerber and Forganni (2021) discerned a growing security discourse in European space policy, with a growing influence of the military-industrial complex in the field and the tentative acceptance, if not active encouragement, by the European Commission. The Commission has been found to be the European institution, which is the most proactive in the security field. Historically, defence has always been one of the most problematic issues of European integration, in which we have found at best not only slow progress but also serious setbacks, such as the rejection of the European Defence Community as early as 1954. Since then, defence policy has not become easier to manage in Europe, and until the Treaty of Maastricht (1992/1993), defence was either left to NATO or to the European nation states, notwithstanding individual initiatives, such as the Franco-German brigade.[1] The Petersberg Tasks were supposed to establish European Defence Policy coming from the Maastricht Treaty. In the Laeken Summit of 2001, the defence capabilities of the EU were still weak to non-existent, which did not prevent the Heads of State and Government to declare the creation of a European Rapid Reaction Force of 60,000 troops deployable within 60 days within 5,000 km of Brussels, and able to be stationed for a year. These objectives have still not been realised more than 20 years later. Back on the bilateral level between the two biggest EU military powers at the time, the St. Malo Agreement between France and Great Britain in 1998 was posited as a major step forward. Nothing much came of it. In 2004, the European Defence Agency was founded in order to arrive at a common defence procurement policy. In general, it has not produced that result.

It has been argued that there are good reasons why defence integration has not worked. The most idealist strand of argument sees the EU as a civilian power based on peace, which would be in contradiction to its own founding principles if it engaged in military policy (Telò, 2006). The case of legitimate defence has of course been made (Legai, 2023: in this book), as a middle ground between pacifism and militarisation, and indeed as an inalienable right guaranteed under international law. Here, however, Europe runs into the divisive question of what should be defended and who has this right. Mostly, this right has been attributed to nation states, as, for example, in the UN charter. Can the EU claim that right, too, as a source of legitimacy for its defence policy?

DOI: 10.4324/9781003230670-1

A more down-to-earth argument has been that every country has historically built up its own military-industrial complex. No government can afford to be seen to lose that national prestige, and the employment and the expertise that could be lost with a shift of that industry away from the nation state to the European level. In its bluntest version, Charles de Gaulle said that 'French defence must be French'[2] (de Gaulle, 1970: 161) and that therefore the link between a nation and its defence cannot be broken. One may add that the military-industrial complex is usually not positively disposed towards sharing its military secrets with other nations. This attitude of secrecy remains well engrained in national defence champions and their engineering researchers (Hoerber, 2006, 2014).

Based on this nationalistic analysis and on the long list of European defence failures, the argument can be made that the dominance of intergovernmental structures in defence policies, such national interests, will always prevail. The European answer to that conclusion is that supranationalism must be introduced. The Commission, as the motor of European integration and therefore perhaps as the most supranational institution, has tried to introduce supranational elements into the defence sector but this has mostly been unsuccessful. Perhaps this is because Europe is no more ready today for a European defence policy than it was in the 1950s. The growing threat to all Member States of the EU, as exemplified in the Russian aggression against Ukraine, tells otherwise in a unified EU response to that aggression. The need for common defence arrangements was shown in the consensual agreement to help Ukraine militarily. Defence, however, remains intergovernmental. Is Europe ready for a supranational defence policy? The reaction to the aggression against Ukraine seems to prove that it is. If that conclusion is true, then that would allow the EU to dispense with what this book has shown space policy to be, the supranational backdoor into intergovernmental defence policy. Before the aggression against Ukraine, Hoerber and Forganni (2022) showed increased military spending in the space sector and thus a growing influence of defence actors in space policy. This might be regrettable, particularly against the backdrop of space having been founded in Europe as a civilian policy with exclusively peaceful purposes (ESA Convention, 1975: Preamble). One may also have the hunch that Europe is growing up, growing out of the shadow of its protector the United States and, by becoming more powerful, also has to be able to defend itself (Hoerber, Forganni, 2022: Conclusion). If Russian aggression ushers in supranational European defence – notwithstanding due cooperation with NATO – then the space sector with its dual-use military and civilian applications will most likely be a staging ground for that European defence, not least because of the military potential in space that has always been there and has only been used sparsely in Europe. We will see the development of these phenomena in the following chapters, before concluding at the end of the book with the inevitably very European question of whether we cannot do better in space than turning it into another war domain. Against the backdrop of the peaceful past of European space policy, there will always be the hope emanating from this continent that peaceful cooperation will take humankind further than confrontation.

Empirical Scope and Methodological Considerations

The end of the Cold War largely removed space from debates on security and defence, since the arms race between the two superpowers – with space being one of the key venues of this race – ended abruptly. However, all hopes for a strictly peaceful use of space dissipated, as new competition for dominance in space quickly kicked off, with major powers reorienting their space policies in an attempt to benefit strategically from the use of space systems for security and military purposes. The EU and key member states have increasingly joined the US, China and Russia, among others, in regarding space assets as critical military, as well as economic, industrial and technological, enablers.

The addition of defence requirements and applications – of an entire multi-faceted military layer – in EU space policy has been the single most important development in the long and fascinating story of the Union's endeavours in space. Up until recently, this addition was mostly taking place in the realm of official discourse, pointing to the need to utilise space for the purposes of the Common Foreign and Defence Policy (CSDP). But today, rhetoric has given way to practise in an empirically verifiable phenomenon that seems to define the very essence and orientation of EU space policy: militarisation. It is true that Galileo and Copernicus, the Union's two flagship space programmes, were indicative of a keen urge to include security and defence in space. As regards Galileo, the inclusion of the public regulated service reserved for EU authorities and member states in the context of the CSDP pointed to a clear non-civilian dimension in satellite navigation. And as far as Copernicus is concerned, its security service geared to support the EU's external action means that the Union has sought to exploit the full spectrum of earth observation's non-civilian applications. But what started as a mostly civilian space programme with some parallel security and defence usages is now turning into primarily a security and defence space programme with secondary civilian applications.

Indeed, the present publication comes in the aftermath of the launch of a fully integrated space programme, spanning the period 2021–2027 and offering a single roof to all EU space activities. The programme is accompanied by a generous financial framework totalling approximately 15 billion euros, intended to fund the development not only of the now mature Galileo and Copernicus programmes but also of the nascent Space Situational Awareness and Government Satellite Communications programmes (Wilson, 2021). The latter two have clear security and defence applications, thereby illustrating the need to engage seriously with an irreversible trend: the involvement of the Union in space for non-civilian purposes. This involvement, however, unfolds in a very complex setting, defined by the co-existence of the European Commission with the European Space Agency (ESA), the EU member states and major space powers that are extremely active in the military and security use of space.

How can this trend towards space militarisation be empirically documented and theoretically analysed? What interests, ideas and institutions are involved in its making in the EU and beyond? How is the drive towards the increased military use of space being promoted politically, facilitated industrially and legitimised discursively? This book tackles these questions by, first of all, shedding light on the military aspects of EU space policy, with special emphasis on the security and defence dimensions of the projects of the European Space Programme. In this context, contributors confront and provide detailed descriptions of the empirical aspects of development, including the role of different institutional actors and the involvement of specific member states. Also, the volume analyses the discursive, ideological, normative and theoretical foundations of the use of space by the EU for strategic purposes, drawing on the broad spectrum of European Integration/ International Relations theory. No single theory has been uniformly adopted by this volume, but an effort has been made by most contributors to base their analysis theoretically and to connect their empirical findings to broader debates concerning the ontology of European integration and/or the international system. Last but not least, the project discusses initiatives outside the EU by key global space players. This reflects our understanding that the European orientation towards a more military-oriented space policy did not appear in a void; on the contrary, it is an episode in a much bigger and broader trend on a global scale.

European space militarisation has been a rather infertile, sporadically covered field of study. There is a reason behind this trend: students of European integration have traditionally viewed space as a mere technical addition to the European project and have disregarded militarisation as a feature that is not compatible with the civilian nature of the Commission's and ESA's engagement, in line with the official rhetoric. In other words, this uneasiness to engage with space and militarisation in the European context stems from the seeming technical complexity of the area: the idea that space is more of a tool rather than a field of politics and power, and the depiction of the EU as a civilian power (Telò, 2006). This, nevertheless, is beginning to change for two main reasons. Firstly, the literature has been enriched by a series of volumes that point to space as a distinct, empirically fascinating and theoretically significant field of study that is of direct relevance to the process of European integration and to European integration theory alike (the role of Routledge's Space Power and Politics series has been key in this regard). Secondly, a series of developments, such as the involvement of the European Defence Agency in space, the maturation of the security applications of Galileo and Copernicus and the introduction of the European Defence Fund, have enabled observers to depart from the official rhetoric and acknowledge the rapid and profound introduction of military purposes and tools in the Union's activities, including those in space.

By overcoming the intellectual fashion of 'civilian space power Europe' and 'space as a technical matter', tremendous opportunities arise for the academic treatment of the Union's key strategic reflections and priorities in the non-civilian use of space, the institutional actors and interests embedded in the generation of these priorities, the military capabilities in space and the political economy of the development of those capabilities. Starting from a well-defined epicentre of the phenomenon of the militarisation of space policy in the EU context, the book covers all of the empirical layers of this phenomenon (member states, European, Global). The chapters document the following claim: the EU is increasingly turning to space for the fulfilment of security and defence purposes and this process has significant political, economic and institutional sources and implications. This shows that the militarisation of EU space policy is part of a broader, global orientation to space as a field of strategic competition. In this respect, despite individual variations, all the chapters are structured around the central themes of the topicality of militarisation being the nucleus of European space policy, the interrelatedness of European security, industrial competitiveness, military capabilities in the shaping of this policy and the way the latter is placed in a broader context beyond the Union.

A conceptual distinction that underpins the present volume should be noted – that of the militarisation versus the weaponisation of outer space. Weaponisation denotes the introduction of space-based systems that act as weapons, i.e., that have a direct destructive capacity. On the contrary, militarisation describes the use of space to support military operations; in this case, the space-based system plays an enabling role in ground-based, air-based and sea-based warfare but does not itself have a destructive capacity (see Mutchler & Venet, 2012: 119; Peoples, 2010: 205–206). Even though the difference is rather subtle, the two concepts maintain their ontological value in order to distinguish between two contexts of the use of space for non-civilian purposes. The development of European space policy discussed in this volume falls under the umbrella of militarisation and not under that of weaponisation; in other words, our use of the term 'militarisation' should not be conflated with the placement of weapons in space.

Structure and Summary of the Chapters

The first section, which includes four chapters, highlights the conceptual aspect of European space militarisation, attempting to both document the relevant processes and present ways of viewing and interpreting these developments. In the first chapter, **Lorna Ryan** argues that the discursive construction of 'the battlefield' is an important element of the process of the militarisation of space. The place where military action occurs is of critical importance in terms of the application of law, specifically that of international humanitarian law. While, as a concept, the idea of a 'battlefield' may not receive explicit attention in international law journals, Megret (2011) refers to how the idea of a battlefield 'haunts' the law. The militarisation of space brings into relief the issue of how the 'battlefield' is constructed, raising questions of what are the 'warfighting domains' and how does this impact on processes of militarisation. This chapter presents the results of a preliminary exploration of EU space policy texts to identify the, possibly shifting, constructions of 'the battlefield'. Its contribution aims to illustrate how the militarisation of space and the weaponisation of space draw, or otherwise, on familiar concepts of 'the battlefield'.

In Chapter 2, **Frank Slijper** observes that space has become a cornerstone of the European Union's security and defence policies, which have grown extensively over the past 20 years. This has been partly driven by technological progress that has generally enabled space assets to become part of our everyday lives, but also because of a new focus on space as a specific, stand-alone domain in military doctrine. In fact, the two developments are interrelated since technological progress is a civilian-military two-way street. However, with more and more countries setting up specific space forces, and with warfare in general becoming increasingly dependent on space assets (navigation, communication, observation), the notion of space for peaceful purposes has gradually eroded, not least after the collapse of the Soviet Union. Moreover, with current developments in the area of emerging technologies (hypersonic missiles, autonomous weapons, and increasingly integrated and automated command architectures) relevant to the space domain as well, renewed rivalry among the United States, China and Russia in particular is affecting the peaceful use of space. While the EU is clearly taking a less assertive approach, it is also afraid to get left behind. This chapter looks into the recent history of the EU's endeavours in the military use of space and analyses policies that are currently guiding it. It, among others, looks into the roles that ESA, EDA, PESCO (Permanent Structured Cooperation) and the newly created European Defence Fund (EDF) play.

Pascal Legai's starting point in Chapter 3 is a wealth of empirical evidence pointing to space militarisation. On 7 September 2018, Madame Florence Parly, French Minister for the Armed Forces, announced the creation of the French Space Command under military governance, similar to the US model, in order to implement a national space strategy. Likewise, NATO decided to declare Space as a full-fledged operational domain at the same level as that of land, marine, air and cyber areas. These essential orientations are the result of the growing willingness of the major space powers, mainly Russia and China, to find dominance by spying from space and using anti-satellite capabilities. The militarisation of space is underway. The EU anticipated this worrying evolution. In this regard, the year 2016 marks a decisive inflexion of the EU towards space development to significantly contribute to the effectiveness of EU external action, including security and defence purposes. The increasing, multifaceted, transnational threats described in the EU Global Strategy for Foreign and Security policy of June 2016 raise the question of the policies to be implemented to tackle them, and importantly about space and related activities. This orientation opens an active debate among the European space actors identified in Article 189 of the Lisbon Treaty: EU, member states and ESA within their respective remits, highlighting in particular the growing role of the European Commission with the Space Strategy for Europe (October 2016) and the European Defence Action Plan (November 2016), and the Regulation establishing the space programme of the Union to be adopted by the European Parliament and the Council in 2020. The current evolution of a stronger Union questions the historic national prerogatives of sovereign responsibility in the security and defence fields that governmental space policies underpin. Thus, European space policy, based on the idea that Security from Space requires Security in Space, sets up a unique space programme encompassing a comprehensive set of dual components: Galileo, Copernicus, SSA/SST, Govsatcom and launchers. It implies the preservation of a world-class research and innovation level, and a state-of-the-art industrial and technological base as part of crucial strategic autonomy. The budgetary proposals for the period 2021–2027 should also remain at a credible level. In addition, the EU did not manage until now to promote a peaceful use of outer space through an internationally recognised legal framework. Within this context, ESA brings its fundamental longstanding experience, despite its Convention excluding defence and military activities, in the design, development and procurement of space systems, in addition to its innovation know-how for downstream applications dealing with huge amounts of space and non-space data. This chapter will elaborate on these different intertwined and sometimes contradictory issues to define an EU future between submission and independence.

Next, in Chapter 4, **Thomas Hoerber** explains the inconsistency of current institutional settings between the ESA and EU space policy, starting from the origin of European space policy and its institutions since WW2. He points out the historical cause of this problematic situation and elaborates on how it undermines potential synergies that were envisaged initially. Based on this analysis, it is proposed to open up ESA internationally, embodied in re-naming ESA to the Space University Institute (SUI), taken from a parallel development of the European University Institute (EUI), or even entertaining the name International Space Agency. It would build on ESA's strength in fostering research, training and innovation, while leaving the utilitarian side of downstream usage of space applications to the EU. This opening up of ESA to a wider international constituency could create partnerships that already exist within ESA with Canada, for example, and with which ESA has substantial experience.

In the second section, there is a move from conceptual clarification to empirical delineation, focusing on concrete policy realms and actors, including the Member States. According to **Isabelle Sourbès-Verger**, in Chapter 5, the issue of space debris has to be considered at the global level, as it is a growing challenge for the international community. Only in the last 20 years, has Europe become increasingly aware of the necessity to deal with the subject and primarily in the goal of securing safety for its satellites. The accumulation of launches since the beginning of the space era – as many as 7,500 objects by January 2017 – has resulted in the multiplication of uncontrolled objects in orbit: launcher stages, defunct satellites, small fragments due to fairing jettison shock and various materials due to processes of deterioration. Even if technical measures were adopted by space agencies to limit its increase, the amount of debris keeps rising, albeit at a moderate rate. The concerns related to space debris and the risks of collision with operational spacecraft are thus increasingly pressing in light of the intensification of space activity, including the expected multiplication of small satellites.

In the next chapter, **Iraklis Oikonomou** claims that, when dissecting the emergence of EU space and defence policies, industrial actors are usually depicted as merely those who are supposed to produce what states and EU institutions ask for in terms of capabilities. In other words, their political role as the drivers of militarisation is not fully appreciated – crushed somewhere among the supranational Commission, the intergovernmental ESA and the myriads of member-state interests. This chapter introduces industrial interests as a key source of policy transformation, examining in particular the role of their Brussels lobbying organisation, Eurospace-ASD, as well as that of individual companies. Adopting a neo-Gramscian perspective inspired by the work of, among others, Bastiaan van Apeldoorn, Andreas Bieler and Kees van der Pijl, the chapter is a first attempt to establish the crucial part played by internationalised space-industrial capital in promoting military space agenda in the EU. This trend is documented empirically via material deriving from interviews, official discourse and publications by the industry. Also, it is interpreted by examining key developments in the global political economy of space production that fuel the industry's quest for more security and defence space programmes at the Brussels level. Overall, it is argued that the European space manufacturers facilitated the emergence of a non-civilian dimension in European space activity, seeing in it a huge opportunity for market expansion and competitive survival.

Turning to the member-state level, **Antonio Calcara** approaches the Italian case starting from the idea that the space sector is at the centre of international competition. Great powers such as the United States, China and Russia look to space as a critical domain for economic, technological and military purposes. The EU has also shown considerable activity, both in supranational policy and at the intergovernmental level of individual states. In the context of this evolving international framework, the chapter focuses specifically on the Italian case study with a twofold objective: firstly, it aims to understand what role space policy militarisation plays in the current Italian security and defence debate, with particular attention to the relationship between the state and the defence and aerospace industries. Secondly, this chapter aims to investigate Italian preferences in the European context. Theoretically, it draws on intergovernmentalism in European integration theory to shed light on the complex relationship between domestic preferences and EU policies in space policy.

Then, **Helen Kavvadia** focuses on a small-state case that of Luxembourg. Thirty years since the end of the Cold War, important developments in the European Union and the international system include radical changes in the security context with new threats, capabilities, 'battlefields' and players. Against this intricate backdrop, and constituting the majority of EU Member States, where do the European small states stand? Do they continue bandwagoning as security consumers following larger Member States, or do they become security contributors? Academic interest has been increasing in both space militarisation and small states. The nexus of the two realms remains under-exposed and under-researched. This chapter instead examines the role of small European states in the militarisation of space, looking at Luxembourg, as a case study. Despite its size, Luxembourg has been a frontrunner in space development. Its enabling environment includes a regulating framework, as well as vibrant technological, financial and academic ecosystems. Building on the dual use of space capabilities, Luxembourg has been expanding into space militarisation applications. Using a structural realist and constructivist perspective, this chapter posits that Luxembourg has been pursuing space militarisation as a window of opportunity for increasing its influence and clout in Europe, as well as in the anarchic structure of the international order.

Finally, the three chapters of the third section offer a glimpse into the global context of space militarisation and, indirectly, into how this may relate to European developments. In Chapter 9, **Mariel Borowitz** turns to the transatlantic dimension of the processes studied in the volume. For years, the U.S. military has maintained the most advanced space surveillance system in the world, monitoring the status of human-made objects in space and providing data and information to spacecraft operators around the world to help avoid collisions in space. In October 2019, the U.S. military announced that it would adopt the term 'Space Domain Awareness' in place of the previously used – and broadly accepted – term 'Space Situational Awareness'. This change was designed to reflect the shift in focus from space as a benign environment to space as a warfighting domain, corresponding with the creation of U.S. Space Command and U.S. Space Force. This chapter examines how both the rhetoric and the substantive approach to Space Situational Awareness/ Space Domain Awareness have shifted in recent years, treating this as an illustrative case study of the trend towards militarisation in recent U.S. Space activities.

In the next chapter, **Dimitrios Stroikos** tackles the role of China and India, arguing that one of the most important aspects of the international politics of space has been the growing use of space assets for military purposes by major space powers, at a time when a wide range of new actors have given impetus to the commercialisation and privatisation of space activities, which renders space an increasingly complex strategic domain. In this reconfigured context, the rise of China and India as space powers has a significant impact on overall space security activities and policies. Indeed, the growing militarisation of the Chinese and Indian space programmes has further consolidated the deterministic notion that conflict in space is inevitable. But while most analyses tend to focus on national security considerations and structural forces that shape the militarisation of space, this chapter suggests that this does not adequately capture the complex array of factors that underlie China's and India's growing interest in the military use of space. In doing so, this chapter opens up with a brief introduction of Sagan's analytical framework on nuclear proliferation that helps offer a more comprehensive approach to the drivers behind the development of military space assets by taking into account national security calculations, domestic politics and the role of state identity. It then moves on to examine the key dynamics that influence China's and India's military space activities as well as their engagement with global space governance from a historical and comparative perspective. It concludes by exploring the implications of China's and India's rise as military space powers for EU space policies and strategies.

Finally, **Jessica West** concludes this volume by highlighting the need for arms control given that, as military competition in outer space rapidly accelerates, a growing number of states are developing both policies and technical capabilities that treat outer space itself as a domain of warfare. Governance has not kept pace. There are few mechanisms in place to prevent crisis escalation and almost none to avoid harmful destruction in this sensitive and critical environment. Focused on the United States and NATO, this chapter reviews the latest military doctrines and strategies to manage the risks posed by this shift towards active warfighting in space, including deterrence, resilience, norms of behaviour and active defences in outer space. Missing from these strategies are formal arms control and other military restrictions in outer space, which this chapter argues are essential to maintaining peace and security in outer space.

Acknowledgements

The editors would like to thank Devon Harvey and Andrew Humphrys at Routledge, Sashivadana at Codemantra and Romain Petrau at ESSCA School of Management for their invaluable contribution to this volume.

Notes

1 See https://www.bundeswehr.de/de/organisation/heer/organisation/10-panzerdivision/deutsch-franzoesische-brigade.
2 original: "(…) la défense de la France soit française.", see also p. 165, see also, Doise, J., Vaisse, M. (1991), *Politique étrangère de la France*, p. 584.

Bibliography

de Gaulle, C. (1970), *Mémoires D'espoir*, Plon, Paris.
ESA Convention (7th ed.) (2010), ESA Communications, Noordwjik.
Franco-German Brigade, https://www.bundeswehr.de/de/organisation/heer/organisation/10-panzerdivision/deutsch-franzoesische-brigade, consulted on 17 March, 2023.
Hoerber, T. (2006), *The Foundations of Europe - European Integration Ideas in France, Germany and Britain in the 1950s*, VS-Verlag, Wiesbaden.
Hoerber, T. (2014), *A Converging Post-War European Discourse - War Experience, Changing Security Concepts and Research & Education*, Lexington books, New York.
Hoerber, T. & Forganni. A. (eds.) (2021), *A Growing Security Discourse in European Space Policy*, Routledge, London.
Hoerber, T. & Forganni, A. (2022), *European Integration and Space Policy – A growing Security Discourse*, Routledge, London.
Jean Doise, Maurice Vaisse, Politique étrangère de la France: Diplomatie et outil militaire 1871–1991, Imprimerie nationale, Paris, 1992.
Legai, P. (2023), 'The militarisation of outer space – A European perspective', in: Hoerber, T. & Oikonomou, I. (eds.), *The Militarisation of European Space Policy*, Routledge, London, pp. 53–65.
Megret, F. (2011), 'War and the Vanishing Battlefield', *Loyola University Chicago International Law Review*, vol. 9, no. 1, pp. 131–155.
Mutchler, M. M. & Venet, C. (2012), 'The European Union as an emerging actor in space security', *Space Policy*, vol. 28, no. 2, pp. 118–124.
Peoples, C. (2010), "The growing 'securitization' of outer space", *Space Policy*, vol. 26, pp. 205–208.
Telò, M. (2006), *Europe: A Civilian Power? European Union, Global Governance*, World Power, Palgrave Macmillan, Basingstoke.
Wilson, A. (2021), 'EU Space programme', Briefing – EU Legislation in Progress, 2021–2027 MFF, European Parliamentary Research Service.

Part I

The Conceptual & Strategic Dimension

1 Militarisation and Space

Constructing Space as Location

Lorna Ryan

Discussion about space as particular type of location has often been addressed in international space law by reference to a concept of 'terra nullius', that is, land that has no owner, itself linked to 'res nullius', that is, a thing that has no owner (see Woodley, 2013: 369). In more general terms, noting that terra is 'land', scholars have considered the implications of the 'empty space' of space between celestial bodies (e.g. Bello y Villarino, 2019). Such discussion gives rise to considerations of sovereignty and to wider considerations of ownership. How space is understood as a particular type of location also holds relevance for a range of social, and political, processes, including the militarisation of space. Drawing on a generally accepted definition of militarisation as 'the use of space-based technology and infrastructure for the purposes of supporting military operations and functions' (Sariak, 2017: 52–53), this chapter considers the designation of space as a location within processes of militarisation. Focusing on European Union space policy, it considers *how* space is discursively constructed as a location in which particular activities occur. While this topic is largely overlooked in the academic literature addressing space policy, the nature of space as a particular location for specific activities is a routine feature of the official texts on European space policy. In January 2022, Thierry Breton, EU Commissioner for the Internal Market, addressing the 14th EU Space Conference, stated, 'space is exponentially a *contested domain*' (European Commission, 2022, emphasis in original). He remarked that space is 'a strategic area where big powers are now competing... Europe must defend its interests and freedom to operate in space. This brings a new strategic dimension to space that must become a strong driver of *all* our plans' (ibid., emphasis added), appealing to collective, rather than individual, interests.

Breton's comments are of interest in situating a discussion about the militarisation of space, indicating the European Union's aspirations as a global actor, as expressed in, for example, *Space Strategy for Europe* (European Commission, 2016) and *Global Strategy for Foreign and Security Policy* (European Union, 2016). Concepts of 'strategic autonomy' and of 'technological sovereignty' are central to discussions of EU goals vis-à-vis space. These features are succinctly captured in a statement in *Space Strategy for Europe:*

> Space capacities are strategically important to civil, commercial, security and defence-related policy objectives. Europe needs to ensure its freedom

DOI: 10.4324/9781003230670-3

of action and autonomy. Space is becoming a more contested and challenged environment. …Growing threats are…emerging in space: from space debris, to cyber threats or the impact of space weather. These changes make greater synergies between civil and defence aspects increasingly relevant. Europe must draw on its assets and use space capacities to meet the security and safety needs of the Member States and the EU…[and] to ensure that Europe maintains autonomous, reliable and cost-effective access to space.

(European Commission, 2016: 8)

The debate as to the nature of these goals, e.g., whether power projection or benign defence, is noted but not addressed in this chapter (for critical analysis, see, for example, Oikonomou, 2013). The current focus is on how, within the EU space policy texts, space is constructed as a location for different activities in the context of the generally accepted understanding of what is termed the dual-use of space. What is the nature of the 'contest' for, in and of the domain of space? The purpose of this question is to direct attention to processes of militarisation. Oikonomou has remarked that militarisation is not pre-determined; 'on the contrary it is a process that can be contested' (2013: 146). Access to, use of and presence in space are central themes in an evolving EU space policy. The justifications for access to, use of and presence in space centre on a concept of 'strategic autonomy':

Space technologies, data and services have become indispensable in the daily lives of European citizens…Space technologies, data and services can support numerous EU policies and key political proprieties…Space is also of strategic importance of Europe. It reinforces Europe's role as a stronger global player and is an asset for its security and defence.

(European Commission, 2016: 2)

The link to strategic autonomy is also evident in the (legal) Regulation establishing the European Space Programme (Regulation 2021/696):

To achieve the objectives of freedom of action, independence and security, it is essential that the Union benefits from an autonomous access to space and is able to use it safely. It is therefore essential that the Union supports autonomous, reliable and cost-effective access to space, especially as regards critical infrastructure and technology, public security and the security of the Union and its Member States. […]To remain competitive in a rapidly evolving market, it is also crucial that the Union continues to have access to modern, efficient and flexible launch infrastructure facilities and benefits from appropriate launch systems.

(European Union, 2021: 70)

That 'space is a…contested domain' directs attention not only to the legislative framework, in particular, the Outer Space Treaty (OST) (1967), with its provisions for the 'peaceful' use of space, but also to the wider discussion of 'dual-use' of space within that framework (itself generally regarded as urgently requiring revision in the context of technological and other developments). It is generally acknowledged that space technology can be used for both military and for civilian purposes, that is, what is meant by 'dual-use':

> [M]ost space technologies, infrastructure and services can serve both civilian and defence objectives. Although some space capabilities have to remain under exclusive national and/or military control, in a number of areas syner-gies between civilian and defence can reduce costs, increase resilience and improve efficiency.
>
> (European Commission, 2016: 10)

The military foundations of space technology are acknowledged (Sariak, 2017); this provenance means for commentators like Sariak that space is already mil-itarised. The literature on the concept and nature of 'militarisation' is consider-able and not reviewed here other than to signal the key texts informing a wider understanding of militarisation; this corpus includes work such as Peoples (2010); Mutschler and Venet (2012); and Bowen (2014, 2019). It is linked to a wider body of literature on security (e.g. Marchisio, 2015) and 'securitisation' (McDonald, 2008; Baker-Beall, 2009; Peoples, 2010; Browning and McDonald, 2011; Beclard, 2013; Sariak, 2017); and on territorialisation and space as a 'global commons' (Elhefnawy, 2003). This literature indicates the dynamic nature of how 'security' and 'security threats' are understood; how they are constructed and the material – that is, real world – effects that may follow from particular understandings of phenomena, for example, constructions of risk and threat, requiring military action (see Ryan, 2020).

The understanding of 'militarisation' used in this chapter is based on the gener-ally used definition as provided by Sariak (2017), cited above, and that provided by Peoples, 'the use of space-based technology and infrastructure for the purpose of supporting military operations' (2010: 76). The related term 'weaponisation' is understood as the actual placement of weapons in outer space (Sariak, 2017: 53). Militarisation is also broadly understood as an activity undertaken to territorialise space (Elhefnawy, 2003) in which 'mere ownership of space systems [is] less of a guarantee of access to space than they once were' (2003: 55):

> states may see 'territorialisation' through the laying of claims to given sec-tions of space as a way of bolstering that control [of space], and the erection of a regime of controlled access as a way of restraining other, more powerful states, otherwise able to dominate the skies above them.
>
> (2003: 57)

Elhefnawy's (2003) comments direct attention to the processual nature of ter-ritorialisation; this same processual nature applies to militarisation. The generally accepted definition of militarisation (Peoples, 2010; Sariak, 2017, above) does not adequately convey the process of militarisation of space – the processes by which the understanding of the access to, use of and presence in space for military purposes is produced and reproduced; within this process, the appeal to the legitimacy of the military presence in and military use of space-based technology is central. The legitimisation is achieved through the justification of the military-related aspects of space-related activity (technology development, application). Considering militarisation as a process, Henry and Natanel query its nature as a fixed condition, pointing to its socially produced nature:

> Militarisation is understood not as a homogeneous and complete exercise of spatial power, but rather as a process which is constantly in flux as well as *continually negotiated*, reiterated and resisted…
>
> (2016: 850, emphasis added)

The 'continual negotiation' includes statement and re-statement(s) of the justification of presence and action in space. It also, crucially, indicates that the spatial aspect is foregrounded in such justifications; the location of areas that are militarised and weaponised. Commentators have suggested that the distinction between weaponisation and militarisation is somewhat artificial. Peoples remarks, 'proponents of the space weaponization make the case the potential dual-use function of non-military satellites already constitutes a de facto form of weaponization' (Peoples, 2010: 205).[1] He succinctly identifies a key issue in 'security studies' as follows:

> The definitional terrain associated with space arms control and the subject of space security more generally is …notoriously fraught… Key points of contention include the question of what constitutes a space weapon or an anti-satellite (ASAT) weapon, the vagaries of the potential 'dual-use' functions of ostensibly non-military space technologies, and the issue of whether a meaningful distinction between the 'militarization' and 'weaponization' of space can be drawn and maintained.
>
> (2010: 205)

He further comments, '[m]ost arms control proponents argue that a distinction can be made between passive systems (militarization) and active systems (weaponization)' (2011: 205).

The foregoing raises an issue, neglected in the literature, as to how space, as a location in which military activity (either through the situation of satellites/in transit action/other) occurs, is constructed.[2] It is notable that, for instance, in a discussion about the diplomacy function achieved by EU research and technological development programmes, Bello y Villarino comments that as a result of military-related action such as cyber-attacks on satellites and methods to jam signals from space:

> we have today a militarised space, where a quarter of the active satellites have some military use. *Space is today a theatre in war plans*. From a legal

point of view, this militarisation was made possible through a particular interpretation of Article IV of the 1967 Outer Space Treaty [which] distinguishes between 'peaceful purposes' applicable to space in general – and 'exclusively peaceful purposes' – restricted to certain celestial bodies. Military uses of the moon and other celestial bodies are then outrightly prohibited, but the 'empty space' between the celestial bodies can be militarized.

(2019: 2, emphasis added)

Bello y Villarino comments on how such developments are 'leading military actors to consider the Earth's orbit a new "warfighting domain"' (2019: 3). He notes, providing examples, that some commentators view these developments not as a race 'to dominate space' but as an incremental development of 'a range of options to control or deny outer space in a time of open conflict' (2019: 3).

Lyall and Larsen remark *a propos* the military use of space that the question remains as to 'what states may do militarily through using space assets or capabilities or space simply [remaining] as a medium' (2018: 453). This question is instructive, pointing to the different ways in which space as location may be conceptualised; as a site in which assets are located or as a transit area. The research question to which this observation gives rise, given EU aspirations relating to access to, use of and presence in space and the overt acknowledgement of the dual-use nature – of space technology, of space access, presence in and use of, space, is *how is space constructed as an object of militarisation in EU space policy?* This research question directs attention to how space *qua* location is constructed as a domain in which particular activities take place, and as a concomitant, the legitimacy of action in a place. Focusing on this dimension of militarisation directs attention to the ways in which justifications for independent access to, uncontrolled presence in and self-determined use of space for military purposes are clearly constructed as 'self-protective actions', defined by Ercan and Kale as 'actions that disrupt the attacker from harming space systems' (2017: 21), and in EU, space policy are set out/enunciated as protection of the EU assets for the European economy. The increasing importance of space has a corresponding sense of the increasing requirement for protection –this is illustrated by Borrell's remark, 'we are [now] very much aware of how important *space* has become as part of the tools that we need to have, master and control in order to ensure our collective security' (EEAS/Borrell, 2022).

Lyall and Larsen note that pre-emptive self-defence 'differs from anticipatory self-defence in that it is triggered not by a specific event, but from a more general apprehension of being attacked' (2018: 452; see also Baker-Beall, 2009). This understanding of pre-emptive self-defence may be expanded to include a general sense of threat to European industry, given the significant technological and industrial benefits of space, and hence to European competitiveness, its autonomy within a global order and its identity. The response to this generalised threat is correspondingly unfocused. The locations of threats and strategies to address these threats raise issues of where action may legitimately take place. These are not fixed.

The structure of the chapter is as follows: further to this introductory section setting out the broad context of the exploration of EU space policy texts, the impetus for the exploration of what Paasi refers to as the social production of spatiality or 'spatial scales' (2001: 9), an outline of how this preliminary exploration of militarisation and place was conceived is presented (see the 'Methodology, Methods and Data' section). This emphasises an initial, unqualified, relationship made between militarisation and war, giving rise to an initial concern with 'the battlefield'. This initial approach was revised as *a priori* and thus unwarranted; however, key sensitising points from the literature relating to 'battlefields' (Blank, 2010; Megret, 2012) are retained. These include the lack of fixity of places as 'battlefields' – a battlefield may be anywhere – and related legal implications for the treatment of both actions and actors in these locations. The important point that militarisation does not necessarily imply war is restated. The legal framework, in particular Articles I and IV of the OST, 1967 is considered including the concept of space as a global commons. In the 'Preliminary Findings' section, the methodology employed in responding to the research question – *how is space constructed as an object of militarisation in EU space policy?* – is presented. This question was prompted by ethnomethodological approaches to place formulations (especially Garfinkel, 1967; Schegloff, 1972; see also Dennis, 2019). Ethnomethodologically informed analyses explore how 'recognisably correct' descriptions of places as locations of particular activities are constructed in text and talk (Ryan, 1996). Location analysis draws attention to what types of activities, 'properly' and expectably, occur within what locations. This work prompted the consideration of how space is constructed as location within European space policy. A qualitative thematic analysis of the official EU space policy texts focused on how space as a place of action is presented in the 'Discussion and Conclusion' section. The preliminary analysis presented suggests two key characteristics of how EU action in space is constructed: (i) different orders of spatiality are present in space; for example, Low Earth Orbit (LEO) versus outer space and (ii) space as an extension of Earth. Illustrative excerpts from the EU space policy corpus are presented in this section. The question of what is gained from a consideration of how space as a site or object of militarisation is then considered (see the 'Discussion and Conclusion' section). It is noted that the need to protect is founded on claiming benefits to the Union – for its role as a global actor, to the competitiveness of its industries and to its citizens. The sense of threat to the European Union, its economy and the lives of citizens in the event of disrupted access and use means that possibility of threat relates to all areas of life – political, economic and social; pre-emptive self-defence is presented as thus warranted. This section also presents concluding remarks.

Militarisation, Battlefields and Zones of Conflict

The study of European space policy texts initially sought to consider whether and how space is considered a 'battlefield'. How a location is, or becomes, identified as 'a battlefield' is a legal, as well as a general concern: as Megret comments, the 'concept of the battlefield has long structured the understanding of war' (2012: 123):

> Defining the battlefield in war is not only a question of militarily deciding where actual battle will occur, nor is it merely a theoretical or doctrinal exercise. Behind these efforts lies a more fundamental struggle to define what constitute a *legitimate* battlefield and, with it, legitimate forms of war.
>
> (2011: 123)

The formulation of the research question arose further to a consideration of how militarisation may, but not necessarily, lead to war. Oikonomou, rejecting the concept of 'securitisation' as an adequate construct to capture the phenomenon of the use of space for military purposes, notes:

> [m]ilitarization "is at the heart of these activities, not only as a process for preparation and conduct of military activities but primarily as a process for the expansion of particular class relations, social forces' interests and forms of power through military means, *irrespective* of the existence of actual war.
>
> (2013: 142, emphasis in original)

This is an important point as the justifications used to defend the introduction of activities or objects that may be used in military exercises include reasoning appealing to a need for preparatory, that is, pre-emptive self-defense, actions. The initial approach was, drawing particularly on the work of Blank (2010) and Megret (2012), focused on locations of conflict, specifically on whether the concept of 'battlefield' was present in the EU policy texts and what this might mean for an understanding of 'militarisation'. This was *a priori* and is not followed – as per Oikonomou's comment, there is no necessary relationship between militarisation and war, thus no necessary relationship between militarisation and 'battlefield'. However, the key points from this literature are retained for current purposes, illustrating, for example, how due to the dual-use nature of space technology and of space, all areas are possible sites of conflict. The 'battlefield' is not a defined area on Earth or in space; it can be any and all areas. Josep Borrell, High Representative of the European Union for Foreign Affairs and Security Policy, speaking at the same conference as Commissioner Breton (cited above), opened his presentation with the following statement:

> I would like to talk about space, the changes taking place in space and what space means for security and defence. I used to say that *outer space is one of the new battlefields*. Cyberspace, outer space and the high seas are the new battlefields of our time. I wish they were not a battlefield, but space is clearly an issue for EU security and defence. First of all space is about technology… *Satellites are there*…They enable us to frame a better response [to disasters]. So space is part of the peaceful life of us all. But space is also increasingly important for geopolitics and for our security and defence. Our freedom of action depends on a safe, secure and autonomous access to space.
>
> (EEAS/Borrell, 2022, no pagination, emphasis added)

He continued:

> Space is getting more crowded…it is also getting more contested. We are seeing more and more examples of an irresponsible and hostile behaviour, and the weaponization of space. Nowadays everything is being weaponised… every aspect of human activity can become a weapon.
>
> (ibid)

This position presents a view of the dynamic nature of European space policy and is aligned with wider academic and policy discussions about the fit-for-purpose status of the OST1967 (as well as the other international agreements in place):

> So, space is becoming increasingly central to European security and the truth is that our current policy framework -what we call the European Union Space Strategy – is from 2016, 6 years ago. And, at that time, it was written focusing mainly on civilian aspects because 6 years ago the idea of space being a battlefield was not as clear as it is today. The 2016 strategy barely touched on the security and defence dimensions of outer space.
>
> (EEAS/Borrell, 2022)

Referring to the newly introduced *A Strategic Compass for Security and Defence* (European Union, 2021a), Borrell remarks that the new EU Space Strategy for Security and Defence 'will get us up to speed with fast-changing developments in outer space and provide us with the instruments we need to defend our citizens and our interests from external threats'. He concluded his speech:

> we [the EU] are [now] very much aware of how important *space has become as part of the tools that we need to have, master and control* in order to ensure our collective security [...]I am sorry to *talk about the satellites as if they were weapons,* but it seems that from the beginning, mankind has been dreaming about space [...]. Every time that humankind has been looking for new paths to go somewhere else, unhappily it has become a *scenario for war.* Let us hope that it will not happen in discovering the path to the stars.
>
> (EEAS/Borrell, 2022, emphasis added)

Borrell's conclusion sees space, or rather, the presence of satellites in space, as part of an arsenal of the EU, and space as a location is both 'scenario for war' and (to refer back to earlier parts in his speech) 'a battlefield'.

The work by Blank (2010) and Megret (2012) was useful in providing key features of the debate about how 'the battlefield' is considered in the academic legal field. Megret presents the following as a definition of a 'battlefield':

> The battlefield is typically an area, limited in space and time, upon which a battle occurs. The battlefield may be created by the *chance encounter* of enemy

troops, but it may also be agreed upon by opposite armies. The battlefield is not a clearly defined space, not even in the most traditional of battles.

(2012: 123, emphasis added)

Thus, what is identified as a battlefield is fleeting and contingent, notwithstanding later commemorations of battlefields. There are no clear boundaries: 'the battlefield is... as much an *idea* as it is a space, and only when one understands the assumptions underlying the *idea* of the battlefield can one understanding how the battlefield has come under threat' (Megret, 2012: 133, emphasis in original).

Thus, while Megret acknowledges the military significance of a battlefield, he cautions 'the battlefield as such does not exist...It is part of a ...construction of reality that allows us to understand certain armed encounters as battles, themselves part of a larger thing called war'. (Megret, 2012: 133). He considers how the law treats activity in spaces identified as battlefields; to simplify his detailed argument, the battlefield is 'an idea and a normative ideal, even as its reality may otherwise be challenged'. Considering technological advances and their impact on the idea of a battlefield as geographically limited (see 2011: 143), he remarks on the 'expansion' of the space. The battlefield, understood broadly, is characteristic of the activity known as war because it tends to be 'the place where many of the markers of war coincide and its existence manifests willingness for direct combat between troops' (2011: 147). Following 9/11, the concept of the battlefield is conceived as globalised and de-territorialised – it is everywhere; as a consequence, 'the] utter disappearance of the battlefield leaves us with very few criteria to determine what sort of activity is going on and, accordingly, what its proper limitations should be' (2011: 151). The implications of the situation in which the concept of battlefield is not defined are that there is the likelihood that force will be used in situations far removed from what the laws of war anticipated. As a result, an effort to reassert the relevance of the concept of battlefield is starting to be heard, i.e., to reassert the battlefield as a 'recognisable and thus legally regulated space' (2011: 151).

Considering the battlefield in relation to space, the issue of the global character of space is evident. However, other aspects also arise that indicate a concept of 'battlefield' in space is fundamentally 'unstable', that is, the dual-use nature of both the presence of and activities in space means that space by its nature is, at the one and same time, a '(potential) battlefield' and another space for non-military use.

As noted, defining the 'battlefield' is important particularly because international law provides for treatment of those in such a space and assesses actions by reference to the location of the action. Blank considers how many contemporary conflicts, in which states fight against non-state actors and terrorist groups, are '*unbounded by sovereign territorial boundaries* ... preferring tactics aimed at civilians often far from any traditionally understood battlefield, can easily confound attempts to use these existing terms effectively' (Blank, 2010: 3, emphasis added).

This lack of fixity, or fluidity, results in 'complex legal conundrums regarding the application of the law to military and counterterrorism operations'

(Blank, 2010: 4). Indeed, the term 'zone of conflict' is preferred, encompassing areas beyond the traditional battlefield. However, while it 'may well have great value' (2010: 5), as a concept, it raises similarly complex legal questions as the concept of 'battlefield'. Thus, the matter of *where* armed conflict can be conducted against terrorist groups is raised. Ancillary questions include 'when and for how long is an area part of a zone of combat' and how far does this designation extend geographically?' (2010: 7). The determination of action in appropriate or warranted location is of concern. Some territorial areas will have stronger connection to an identified 'zone of combat' than others (Blank, 2010: 38). Blank suggests that identifying '*where* and *when* a state can conduct operations within an armed conflict framework [is] a necessary companion to the ongoing debate about *whether* a state can conduct operations within such a framework' (2010: 38, emphasis in original). The legal order relating to conflict within designated areas is instructive in indicating (a) the fluidity of locations as conflict locations and (b) the resulting legal lacunae arising.

The Legal Framework

The research question, *how is space constructed as an object of militarisation in EU space policy?*, directs attention to how space *qua* location is constructed as a domain in which particular activities take place, and as a concomitant, the legitimacy of action in a place. A key reference point is the legal framework within which such assessments are made. This legal framework is not only the legal framework relating to conflict as indicated in the discussion about the nature of the battlefield but more directly relates to the body of law referred to as 'space law' (see Diederiks-Verschoor and Kopal, 2008; Lyall and Larsen, 2018). Academic law and policy discussions of militarisation – and weaponisation – of space routinely cite the OST 1967 in which Articles 1–4 (see Table 1.1) are particularly relevant to such discussion. This Treaty, and the wider legal framework (five Treaties are generally understood as constituting the legal framework relating to space) constitute the key reference point situating activities in outer space, including the moon and other celestial bodies, in relation to the peaceful purposes of space exploration and use. The principles of non-appropriation, the prohibition on claims by means of occupation or otherwise, are established. The statement that outer space 'shall be the providence of all mankind', i.e., that it is a global commons – open to all, for use by all – is central to the 1967 OST (see below). The use of space for the benefit of citizens as per the *Global Strategy for the EU's Foreign and Security Policy* (European Union Global Strategy/EUGS) draws on the concept of global commons:

> The EU will advance the prosperity of its people… A prosperous Union also hinges on an open and fair international economic system and sustainable access to the global commons.
>
> (European Union, 2016: 8)

Articles I–IV of the OST are presented below. Article IV emphasises peaceful use; this is the 'core' element around which discussions about militarisation are focused.

Table 1.1 Outer Space Treaty 1967 – Key provisions (Article 1 -IV)

Art I	The exploration and use of outer space, including the moon and other celestial bodies, shall be carried out for the benefit and in the interests of all countries, irrespective of their degree of economic or scientific development, and shall be the province of all mankind.
	Outer space, including the moon and other celestial bodies, shall be free for exploration and use by all States without discrimination of any kind, on a basis of equality and in accordance with international law, and there shall be free access to all areas of celestial bodies.There shall be freedom of scientific investigation in outer space, including the moon and other celestial bodies, and States shall facilitate and encourage international co-operation in such investigation.
Art II	Outer space, including the moon and other celestial bodies, is not subject to national appropriation by claim of sovereignty, by means of use or occupation, or by any other means.
Art III	States Parties to the Treaty shall carry on activities in the exploration and use of outer space, including the moon and other celestial bodies, in accordance with international law, including the Charter of the United Nations, in the interest of maintaining international peace and security and promoting international co-operation and understanding.
Art IV	States Parties to the Treaty undertake not to place in orbit around the earth any objects carrying nuclear weapons or any other kinds of weapons of mass destruction, install such weapons on celestial bodies, or station such weapons in outer space in any other manner.
	The moon and other celestial bodies shall be used by all States Parties to the Treaty exclusively for peaceful purposes. The establishment of military bases, installations and fortifications, the testing of any type of weapons and the conduct of military manoeuvres on celestial bodies shall be forbidden. The use of military personnel for scientific research or for any other peaceful purposes shall not be prohibited. The use of any equipment or facility necessary for peaceful exploration of the moon and other celestial bodies shall also not be prohibited.

Methodology, Methods and Data

The methodology employed to explore the research question is qualitative, exploring one aspect of how militarisation is constructed in official texts of the space policy of the European Union. It is (loosely) informed by ethnomethodo-logical approaches to the study of descriptions of social life as accomplishments. In particular, the work on place formulations as interaction achievements is of relevance. This focuses on how recognisably correct descriptions of places as sites of particular activities are produced. It considers how space is constructed as an object of militarisation in official discourse. The data sources are the official texts of the European Union relating to space policy. The data are excerpts from the official texts from 1979 to 2021; the corpus contains documents from the European Parliament Resolution on Space Policy, 1979, to the European Union Regulation establishing the European Space Programme in 2021, a legal text. The research question was refined over the course of this reading, moving from an initial concern with whether and how space is constructed as a battlefield to a consideration of how space as a location of action is constructed in the policy texts. The preliminary focus is to identify the lexicon of location; how the place of action was 'spoken about' in the policy texts. Texts were read to identify themes across the entire corpus. The dataset that was created includes excerpts from the texts, identified as relevant to this research question. The purpose is not to consider the veracity or otherwise of statements but rather to try to explore the production of meaning, to explore how the place of 'militarisation' is constructed in the texts and how 'space' is employed in discussions about dual-use of place, in this case, space.

Preliminary Findings

This preliminary study was prompted, as briefly described above, by ethnomethodology (see Garfinkel, 1967), in particular, its work place formulations as members' accomplishments, rather than pre-existing descriptions of locations (see Schegloff, 1972). It was more directly prompted by considerations of the justification for defence as articulated in EU space policy and related law (European Union, 2021). The research question is *how is space constructed as an object of militarisation in EU space policy?* The thematic analysis suggested firstly that 'space' is constructed in a variety of different ways and secondly that it is presented as an extension of Earth.

The initial approach arose from literature on battlefields, particularly Megret (2012) and Blank (2010). However, a preliminary examination of the empirical data, the EU space policy texts, indicated no references to 'battlefields' or 'zones of conflict'. In this regard, Borrell's speech (EEAS/Borrell, 2022) referred to above is extraordinary in its repeated identification of space as a battlefield. Assuming a link between militarisation and war (and battlefield) was erroneous. However, the preliminary review of the corpus of material – the EU space policy texts from 1979 to 2021 – suggested that the rationale of EU action in support of EU goals, routinely asserted, could be an analytically productive, illustrating how an understanding of the use of place, in this case 'space' broadly defined, is linked to these objectives.

A typical statement about EU goals in space is evident in an excerpt from a Commission Communication, *Europe and Space: Turning to a New Chapter*. This communication refers to how economic, societal and political factors are intertwined: ensuring satellite-based services is identified as 'a political factor for security as well as an instrument of global influence, since an independent satellite capability ensures control over the use of the information gathered' (European

Commission, 2000: 8). This 'political factor' concerns the EU's aspirations to function as a global actor. Securing independent access to space is about defending access for the purpose of non-dependence. It may be linked to the concept of the global commons; a resource for all – the key space Treaties – seeks to maintain this 'open' space.

Constructions of Space as a Place

The lexicon or vocabulary of space as location includes the place terms that are used. These are considered in relation to the broad question 'what are the 'geographies of space'? An example of how 'space' is constructed is as follows:

> Space is often seen as the last *frontier* of mankind's curiosity; an *area* of pioneering technology development associated with science, exploration, defence and informational prestige.
>
> (European Commission, 2000: 6, emphasis added)

Designating different areas of space, Elhefnawy comments:

> [T]he most widely accepted definition appears to be that it is the lowest perigee attainable by an orbiting vehicle; the definition relies less on a system's altitude than the class of system in question and all satellites therefore enjoy complete freedom to overfly the Earth.
>
> (2003: 60)

Elhefnawy notes the absence of a formal definition of space (ibid). This definitional void is of note – where space *is* not clear – outer space is marked by the Karman line; this represents an attempt to set a marker between the Earth atmosphere and space – 100 KM above the planet's surface (see also Diederiks-Verschoor and Kopal, 2008: 23 ff and McDowell, 2020: 1 for a distinction between upper and lower Low Earth Orbit, for example).

In the literature, Elhefnawy comments, 'the conventional wisdom is that space will remain a commons' (2003: 55), a perspective that is rooted in what he terms the 'profound ambiguities inhering in international law over the enforcement of the regime for the use of space, and indeed, even where the boundary of space is' (2003: 55). Considering territorialisation as a process relating to controlling access, he foregrounds processes of discrimination between different areas:

> Movement towards a regime of 'controlled access' at sea [e.g. Exclusive Economic Zones] is a reminder that states are fundamental territorial entities, with a tendency towards expanding their sphere of responsibility, even when they cannot exercise effective control within that sphere, a tendency that can be expected to carry over into space.
>
> (2003: 56)

Further, considering the notion of a zero-sum feature of access by reference to the limited number of geostationary slots and the bandwidth usable by satellite communications to illustrate his point, he comments:

> Even without states attempting to control such slots access to space is coming to be regarded less as a question of owning space systems than of militarily exercising control of space itself...Ownership of satellites takes a backseat, while the ability to protect one's satellites or attack those of an enemy combine to make control of *strategically important portions of space* important.
>
> (2003: 57, emphasis added)

That different areas of space into those that are strategically important or unimportant are foregrounded as being more or less important is of interest, pointing to a nuanced understanding of space as a particular type of location that is internally differentiated. The militarisation of space is about the use of space-based assets for Earth-based actions; however, activities also occur in space and assets/objects may use space as a route. In such scenario, space is a medium. The different characterisations of space hold relevance for the definition of militarisation as the use of space-based technology and infrastructure for the purpose of supporting military operations (see Peoples, ibid). The focus of attention is not only about the placement of objects in space but also about how space is viewed.[3]

In 1988, in the first Communication from the Commission, space is described as:

> *an area* in which the Community now seems destined to play a broader and more active role...The era of the conquest of space has given way to an era of space exploitation.
>
> (European Commission, 1988: 1, emphasis added)

In 2001, the European Parliament described space as '*a platform* for military action' (2001: 13, emphasis added). To some extent, this is the end point of militarisation – in this instance, space is conceptualised as the place from which action is launched:

> The Space Advisory Group in 2012, commented that '[t]here is a need for consolidated, shared vision for robotic and human exploration of Mars, the Moon and near Earth objects... Space should be a tool of the EU for international cooperation...'
>
> (2012: ixx)

The militarisation of space can be presented as affecting different 'parts' of space – from outer space to LEO. Within the space policy texts, references to space as 'a platform', 'tool', 'segment', as well as the specific references to space as a location of action, are of note. This is because without specificity, 'space' may be used to denote support to (military) activities as well as the location of such activities. It is constructed as an 'empty space' between celestial bodies. Bello y Villarino regards the militarisation of space as a result of the interpretation of the provisions of the OST1967 in which the distinction between peaceful purposes applicable to space generally and 'exclusively peaceful purposes' applicable to the Moon and celestial bodies (2019: 2, see Note 1).

In the EU Global Strategy for the EU's Foreign and Security Policy, the following statement indicates this understanding of space as a differentiated place:

> We [the EU] have an interest in fair and open markets, in shaping global economic and environmental rules, and in sustainable access to the global commons through open sea, land, air and *space routes.* In view of the digital revolution, our prosperity also depends on the free flow of information and global value chains facilitated by a free and secure Internet.
>
> (2016: 15, emphasis added)

The use of space for the purposes of transit, that is, space as a medium (Lyall and Larsen, 2018) as well as a specific resource – the global commons – is suggested here. As Fluri has suggestively remarked: '[t]axonomies of...spaces are necessary for representations of state security in order to frame epistemologically what is 'secure' and what is not.. [Contemporary state] security therefore reveals interrelated political, social and economic parameters' (2014: 797).

This interrelationship is evident in the EU's 2021 Regulation in which the statement that 'the possibilities that space offers for the security of the Union and its Member States should be exploited' is followed by an elaboration of what this means:

> Historically, the space sector's development has been linked to security. In many cases, *the equipment, components and instruments used in the space sector, as well as space data and services, are dual-use.* However, the Union's security and defence policy is determined within the framework of the Common Foreign and Security Policy, in accordance with Title V of the Treaty on European Union (TEU).
>
> (2021: paragraph 2, emphasis added)

Space as an Extension of Earth

Sariak (2017) reiterates Johannes Wolff's position that military space technology operates as a 'force multiplier', amplifying the effects of older, more conventional forces. He suggests that this indicates how 'those who advocate a weaponization of space conversely see space as a theatre for conflict itself, rather than the ultimate high ground for operations on Earth' (2017: 53). Bowen has similarly commented, '[s]pace warfare is the continuation of Terran [Earth] politics by other means, and the command of space connects space warfare to those wider political goals' (2019: 540). This perspective is suggestive in terms of how the militarisation of space 'connecting' space to Earth may be understood in terms of Earthly goals. These goals, that is, autonomous access to, use of and presence in space, are not for space dominance *per se* but are for the protection of the Earth (that is, 'pre-emptive self-defence' as discussed by Lyall and Larsen, 2018, above). This theme is prominent in the EU space policy texts as the following excerpts illustrate:

> space has contributed in the past and will undoubtedly continue to contribute in the future to the building of Europe: it serves both as a focus for the European identity, a frontier which has a mobilising influence on technology and the economy and a measure of political power.
>
> (The Community and Space: A Coherent Approach,
> European Commission, 1988: 10)

How space can be used is a recurrent theme within the policy texts. In an early Communication on space policy, the European Commission noted a key question:

> [...].An *integrated Europe* with 400 million citizens and a land area of 4 million km [squared] must ask the question of *where it wants to position itself in space* and space applications and whether it wants to develop a vision for the future of its industry and service providers to preserve at least a dual-source situation in the world in order to ensure that competitive procurement remains an option.
>
> (1999: 5–6, emphasis added)

The 'Earthly' gains arising from space dominance are clear. In 2000, the Council Resolution on a European Space Strategy reaffirmed 'the strategic nature of space' and the need 'to conduct an overall space policy reflecting Member States' political ambitions and responding to the challenges of European integration' (European Council, 2000: 1). The alignment with wider EU goals is evident in the Communication *Towards a European Research Area* (European Commission, 2000), in which the following characterisation is presented:

> *Global information and communications constitute the nervous system of the knowledge society. Satellites, with their ability to cover and to connect virtually every point around the world, are critical to the effective functioning of this neural network.* The observation satellite systems deliver a continuous flow of near real-time data about any part of the globe, in compliance with international law. This is of vital importance...for early warning of crises and for arms control...Satellite based services are of strategic value to Europe, where economic, societal and political factors are inseparably mixed...Galileo...will provide Europe with sovereignty in safety critical applications and telematic infrastructure.
>
> (2000: 7, 14, emphasis added)

Satellites in space form a link to the 'nervous system' of Earth; they become extensions of this system. In *Space Strategy for Europe*, the Commission noted the ways in which space technologies data and services can – and do – support numerous EU policies and key political priorities: 'Space is... of strategic importance for

Europe. It reinforces Europe's role as a stronger global player and is an asset for its security and defence' (European Commission, 2016: 2).

Finally, in the recital of the 2021 *Regulation establishing the European Space Programme*, a legal text, the dual-use of systems and their applications are presented as a matter of fact:

> Owing to the importance of space-related activities for the Union economy and the lives of Union citizens, the dual-use nature of the systems and of the applications based on those systems, achieving and maintaining a high degree of security should be a key priority for the Programme, particularly in order to safeguard the interests of the Union and of its Member States, including in relation to classified and other sensitive non-classified information.
>
> (European Union, 2021, point 51).

The common sense enunciated in this instance is that space systems and presence in space are simultaneously for military use and purpose *and* for civil use and purpose. Satellites may be used for either, or both, purposes. The contestation is removed – systems *are* 'dual-use'.

Discussion and Conclusion

What has been gained from a consideration of how the official texts address space within the wider context of processes of militarisation? The answer to the research question, *how is space constructed as an object of militarisation in EU space policy*, is that space is variously constructed as both 'a tool' or 'instrument' and 'a platform' from which, to attain a strategic position, as well as a particular location within which activities can occur and objects can be placed. This finding holds relevance for an understanding of how the process of militarisation has a ubiquitous reach.

As a process of militarisation, that is, the progressive infusion of military requirements, needs and considerations, within a specific policy field, the location of military activity, 'the battlefield', with associated rules of action and benefits/penalties accruing, may be ascribed a constitutive role in processes of 'militarisation'. In 2010, Peoples commented, 'outer space is becoming ever more 'securitized'; that is, *access* to space is now commonly framed as crucial to the military, economic and environmental security of leading states and international organisations' (2010: 205, emphasis added). Over a decade later, this goal of access to (alongside presence in) space has extended to include use of space – the EU needs access to, presence in (via its assets) and use of space for its economy and security. This supports Lyall and Larsen's assessment (presented in the context of US activity in space), 'the line between military and civilian-political concerns might be increasing difficult to discern' (2018: 475).

While within space policy security as an issue has increased in prominence in the last 40 years of European Union space policy formulation (see European Parliamentary Research Service, 2017), the harnessing of space policy explicitly to the Common Security and Defence Policy and within the Global Strategy for Foreign and Security Policy (European Union, 2016) is a recent phenomenon and one that is likely to continue. The six-year period between the EU's Space Strategy for Europe (2016) and the 14th European Space Conference (January 2022) is characterised by significant developments in relation to security – in this view, the recognition of space as a battlefield, a pre-existing reality that was not visible, becomes visible (see EEAS/Borrel, 2022[4]). Borrell's candour about the militarisation process is striking, instancing what Hoerber and Forganni have referred to how:

> Security has become a more important issue in space circles. There is still a major hesitation to call it defence, militarisation or even weaponisation, all terms used in the past ten to fifteen years to describe military influence on the space sector, which in a European context was founded as a civilian endeavour for peaceful purposes.
>
> (2020: 14)

Considering why such hesitation exists is more than a matter of academic concern, they state that:

> [A] more profound answer to the question why we are talking about security and defence issues in the European sector, today, is that the world is perceived to have become a more dangerous place; and that the EU in particular has got to a stage of influence in the world and institutional development where it has to protect its assets.
>
> (ibid)

The process of the militarisation of space is presented as unavoidable, a consequence of dual-use technology; co-presence of military and civil activity is inevitable: the recital in Regulation 696/2021 remarks '[h]istorically, the space sector's development has been linked to security. In many cases, the equipment, components and instruments used in the space sector, as well as space data and services, are dual-use' (European Union, 2021). The OST 1967 is assessed as outdated in terms of contemporary technological developments and while the use of space for 'peaceful' purposes may be seen as routinely constituting a reference point to discussions about the use of space, it was, remarkably, not referenced in the current *Space Strategy for Europe* (European Commission, 2016). The analysis presented indicates an increasing foregrounding of the rationale for militarisation, that is, the justification(s) of the use of space-based systems and space infrastructure – including ground segments –for pre-emptive self-defence purposes is routinely mentioned in official texts on space. Space is a tool for EU aspirations – political, economic and social. Megret's (2012) comment about the 'legitimate battlefield' directs attention to appeals for legitimacy of actions; specifically, these appeals relate to the importance of space for the economy, the lives of EU citizens and the capacity of the EU to act on the global stage.

The policy texts function as discursive sites of contestation: how space is understood, either as a tool or as a possible site or location for conflict will affect how the institutions of the EU and the associated policies are designed and implemented. Considering the constant 'process of negotiation' (Henry and Natanel, 2016) directs attention to contestation within policies. How space is understood as a location is central to this process of negotiation.

Detailing a proposed future research agenda for space policy, Schrogl et al. comment, 'the militarization of outer space is a standing issue... As most space policy is dominated by law, natural science and engineering, bringing in social sciences could enhance our perception of militarisation, weaponization and securitization' (2021: xxxviii). This exploratory study of how space is constructed in a particular discourse site, EU space policy texts, suggests that an understanding of how 'places' in space are constructed, or more precisely, how space is constructed as a location in which particular activities, for particular purposes, take place, would provide a much-needed contribution to the wider research agenda.

Acknowledgements

Thanks to Iraklis Oikonomou and Thomas Hoerber for constructive comments on an earlier version of this chapter.

Notes

1 Oikonomou foregrounds the wider context of EU space policy militarisation: it "belongs to a broader context of EU militarisation per se, defined as 'the contradictory and tense social process in which civil society organises itself for the production of violence'" (Gills, cited in Oikonomou, 2013: 134).
2 There is a rich literature on how space has been conceptualised, the "space imaginary" (see Shukaitis, 2009), this is not addressed here but is noted as part of a contextual web within which policy texts are constructed and received.
3 The ethics of the different perspectives are not discussed here; suffice it to note that this constitutes a separate area for consideration (see for example, Williamson, 2003 for an overview of key ethical issues).
4 It is a suggestive comment in its intimation that space has always been a battlefield, it was just that it was not recognised as such ('the idea of space being a battlefield was not as clear as it is today'). Megret (2012) has suggested the concept of the battlefield '*haunts*' the law; this preliminary exploration suggests the same may be said for European space policy: the widening process of space securitisation, as commented on by scholars such as Bowen (2014), has the effect that the 'battlefield' of space is potentially always possible; access to, use of and presence in space are at one and the same time military and civil.

Bibliography

Baker- Beall, C. (2009) 'The discursive construction of EU counter-terrorism policy: Writing the 'migrant other', securitisation and control', *Journal of Contemporary European Research*, 5(2): 188–206.
Beclard, J. (2013) '"With the head in the air and the feet on the ground": The EU's actorness in international space governance', *Global Governance*, 19: 463–479.

Bello y Villarino, J-M. (2019) 'Preventing a cold war in space using European research and innovation programs', *Science and Diplomacy*, 8(1): 1–12. https://www.sciencediplomacy. org/article/2019/preventing-cold-war-in-space-using-european-research-and-innovation- programs <accessed 18.2.2022>

Blank, L. (2010) 'Defining the battlefield in contemporary conflict and counter-terrorism: Understanding the parameters of the zone of combat', *Georgia Journal of International and Comparative Law*, 39(1): 1–38.

Bowen, B.E. (2014) 'Cascading crises: Orbital debris and the widening of space security', *Astropolitics*, 12(1): 46–60.

Bowen, B.E. (2019) 'From the sea to outer space: The command of space as the foundation of spacepower theory', *Journal of Strategic Studies*, 42: 3–4, 532–556.

Browning, C. and McDonald, M. (2011) 'The future of critical security studies: Ethics and the politics of security', *European Journal of International Relations*, 19(2): 235–255.

Dennis, A. (2019) 'The influence of 'topic and resource' on some aspects of social theoris- ing', *Journal for the Theory of Social Behaviour*, 49(3): 282–297.

Diederiks-Verschoor, I.H. and Kopal, V. (2008) *An Introduction to Space Law*, 3rd ed. Alphen aan den Rijn: Wolters Kluwer.

Elhefnawy, N. (2003) 'Territorializing space? Revising an old idea', *Astropolitics*, 1(2): 55–63.

Ercan, C. and Kale. O. (2017) 'The role of space in the security and defence policy of Turkey: A change in outlook: Security in space versus security from space', *Space Policy*, 42: 17–25.

European Commission. (1988) *The Community and Space*. COM (88) 417 final.

European Commission. (1992) *The European Community and Space: Challenges, Opportu- nities and Actions*. COM (92) 360 final.

European Commission. (1996) *The European Union and Space: Fostering Applications, Markets and Industrial Competitiveness*. COM (96) 617 final.

European Commission. (1999) *Towards a Coherent European Approach for Space*. SEC (1999) 789 final.

European Commission. (2000) *Europe and Space: Turning to a New Chapter*. COM (2000) 597 final.

European Commission. (2003) *Space: A New European Frontier for an Expanding Union: An Action Plan for Implementing the European Space Strategy*. COM (2003) 673 final/ SEC (2003) 1249.

European Commission. (2005) *European Space Policy – Preliminary Elements*. COM (2005) 208 final.

European Commission. (2007a) *European Space Policy*. COM (2007) 212 final.

European Commission. (2007b) *Summary of the Impact Assessment of the European Space Policy*. SEC (2007) 506.

European Commission. (2011) *Towards a Space Strategy of the European Union That Benefits Its Citizens*. COM (2011) 152.

European Commission. (2013) *Establishing a Space Surveillance and Tracking Support Programme*. Proposal, COM (2013)107 final.

European Commission. (2016) *Space Strategy for Europe*. COM (2016) 705.

European Commission. (2018a) *Establishing the Space Programme of the Union and the European Agency for the Space Programme*. COM (2018) 447 final.

European Commission. (2018b) *Space and Security*. https://ec.europa.eu/growth/sectors/ space/security_en <accessed 20.1.2020>

European Commission. (2022) *Speech by Commissioner Breton at the 14th Space Conference.* (europa.eu) <accessed 6.5.2022>

European External Action Service/EEAS. (2022) *Space: Speech by High Representative/ Vice President Josep Borrell at the 14th European Conference\EEAS.* Website (europa. eu) <accessed 6.5.2022>

European Parliament. (1979) *Community Participation in Space Research.* C 127/24, 21.5.79.

European Parliament. (1998) *Resolution on the European Union and Space.* OJ C 34, pp. 27–30, 2.2.1998.

European Parliament. (2001) *Report on the Commission Communication to the Council and the European Parliament on Europe and Space: Turning to a New Chapter* (COM (2000)597-C5–0146/2001/2072 (COS)), Committee in Industry, External Trade, Research and Energy, PE309.051.

European Parliament. (2013a) *Report on EU Space Industrial Policy, Releasing the Potential for Growth in the Space Sector*, Committee on Industry, Research and Energy, PE514.925v02-00.

European Parliament. (2013b) *Opinion of the Committee on Foreign Affairs for the Committee on Industry, Research and Energy on EU Space Industrial Policy.* Releasing the Potential for Growth in the Space Sector PE514.674v02-00.

European Parliamentary Research Service. (2016) *European Space Policy: Industry, Security and Defence.* PE583.790.

European Parliamentary Research Service. (2017) *European Space Policy: Historical Perspective, Specific Aspects and Key Challenges.* EPRS/Reillon, V. PE 595.917.

European Union. (2016) Shared Vision, Common Action: A Stronger Europe. *A Global Strategy for the European Union's Foreign and Security Policy.* EU Global Strategy | EEAS Website (europa.eu)< accessed 6.5.2022>

European Union. (2021a) *Regulation (EU) Establishing the Union Space Programme and the European Union Agency for the Space Programme.* Regulation (EU)2021/696.

European Union. (2021b) *A Strategic Compass for the European Union.* https://www. eeas.europa.eu/sites/default/files/documents/strategic_compass_en3_web.pdf <accessed 6.5.2022>

Fluri, J. (2014) 'States of (in)security: Corporeal geographies and the elsewhere war', *Environment and Planning: D: Society and Space* 32: 795–814.

Garfinkel, H. (1967) *Studies in Ethnomethodology*, 11th ed. Cambridge: Polity Press in association with Blackwell Publishing.

Henry, M. and Natanel, K. (2016) 'Militarisation as diffusion: The politics of gender, space and the everyday', *Gender, Place & Culture*, 23: 850–856.

Hoerber, T. and Forganni, A. (2020) 'Introduction', in Hoerber, T. and Forganni, A. (Eds.) *European Integration and Space Policy: A Growing Security Discourse*, 1–19. London: Routledge.

Lyall, F. and Larsen, P. (2018) 'The military uses of outer space', in Lyall, F. and Larsen, P. (Eds.) *Space Law: A Treatise*, 447–481. London: Routledge

Marchisio, S. (2015) 'Security in space: Issues at stake', *Space Policy*, 33(Part 2): 67–69.

McDonald, M. (2008) 'Securitization and the construction of security', *European Journal of International Relations*, 14(4): 563–587.

McDowell, J. (2020) 'The Low Earth Orbit satellite population and impacts of the SpaceX Starlink constellation', *The Astrophysical Journal Letters,* 892: L36, 10.

Megret, F. (2012) 'War and the vanishing battlefield', *Loyola University Chicago International Law Review*, 9(1): 131–155.

Mutschler, M. and Venet, C. (2012) 'The European Union as an emerging actor in space security?' *Space Policy*, 28: 118–124.

Oikonomou, I. (2013) 'The political economy of EU space policy militarisation', in Stavrianakis, A. and Selby J. (Eds.) *Militarism and International Relations: Political Economy, Security, Theory*, 133–146. London: Routledge.

Paasi, A. (2001) 'Europe as a social process and discourse: Considerations of place, boundaries and identity', *European Union and Regional Studies*, 8(1): 7–28.

Peoples, C. (2010) 'The growing "securitisation" of outer space', *Space Policy*, 26: 205–208.

Ryan, L. (1996) *'Reading the Prostitute': Appearance, place and time in British and Irish Press Stories of Prostitution*. Aldershot: Ashgate.

Ryan, L. (2020) 'Security and the discourse of risk in European space policy', in Hoerber, T. and Forganni, A. (Eds.) *European Integration and Space Policy: A Growing Security Discourse*, 75–96. London: Routledge.

Sariak, G. (2017) 'Between a rocket and a hard place" military space technology and stability in international relations', *Astropolitics*, 15(1): 51–64.

Schegloff, E. (1972) 'Notes on a conversational practice: Formulating place', in Sudnow, D. (Ed.) *Studies in Social Interaction*, 79–118. New York: The Free Press.

Schrogl, K-U., Giannopapa, C. and Antoni, N. (2021) *A Research Agenda for Space Policy*. Cheltenham: Edward Elgar Publishing.

Shukaitis, S. (2009) 'Space is the (non)place: Martians, Marxists and the outer space of the radical imagination', *Sociological Review 57 (s1) Special Issue: Sociological Review Monograph Series: Space Travel & Culture: From Apollo to Space Tourism*, edited by David Bell and Martin Parker, 98–113.

Williamson, M. (2003) 'Space ethics and the protection of the space environment', *Space Policy*, 19(1): 47–52.

Woodley, M. (ed) (2013) *Osborn's Concise Law Dictionary*. London: Sweet and Maxwell.

2   Europe's "Defensive Militarisation" of Space

Space in the Context of the EU's Emerging Military Agenda

Frank Slijper

Introduction

In presenting the European Union's draft Strategic Compass in November 2021, the EU High Representative for Foreign Affairs and Security Policy, Josep Borrell, poses:

> Europe is in danger: we need to operate in an increasingly competitive strategic environment. The purpose of the Strategic Compass is to draw an assessment of the threats and challenges we face and propose operational guidelines to enable the European Union to become a security provider for its citizens, protecting its values and interests. […] In recent years, the classic distinction between war and peace has been diminishing. The world is full of hybrid situations where we face intermediate dynamics of competition, intimidation and coercion. […] The geopolitical stage is also becoming more complex. More and more states are behaving as partners on certain issues and competitors or rivals on others. International relations are increasingly organised on a transactional basis. For the EU, which remains the world's most open space and which borders many areas of conflict, this is a real challenge. Even more so because there are also worrying dynamics such as the collapse of states, the retreat of democratic freedoms, violations of international and humanitarian law, plus the attacks on the 'global commons': cyber space, the high seas and outer-space.
>
> (EEAS, 2021)

It is unclear to what Borrell exactly refers to when mentioning *attacks on* outer space. He must mean either real attacks by space powers (US, China and Russia) on their own space assets in the form of anti-satellite (ASAT) missile tests (BBC, 2008; West, 2020a; Gohd, 2021), or the general risk of space debris hitting space assets, or he may have meant the potential of attacks, including through cyberattacks, to satellites or supporting ground systems. Regardless, it is clear that space has become a domain of key importance to the functioning of much of our world, and that any threat to that could lead to major trouble.

DOI: 10.4324/9781003230670-4

This chapter will explore the roots and evolution of the European Union's space policy, and its military dimensions in particular. It will first look into how it emerged in the early years of this century. It will then look at how these relate to parallel developments in agencies such as NATO, the European Space Agency (ESA) and the European Defence Agency (EDA) and analyse how space policy and practice have developed over time to become a cornerstone of the EU's security and defence policies. Finally, it concludes that while it is logical for the EU to plan and prepare against attacks – which the EU calls "defensive militarisation of space" – it is also clear that it will not stop an arms race in space. For that, Europe should use its diplomatic clout more and step up its role in multilateral initiatives.

EU

Although the 2003 EU Security Strategy – the EU's first – also refers to constantly evolving, more diverse, less visible and less predictable threats, the general tone is less alarmist, while against the background of the ongoing US-led "global war on terror" in response to the September 11, 2001 terrorist attacks, and the 2003 Iraq war that had just started: "Europe has never been so prosperous, so secure nor so free. The violence of the first half of the 20th Century has given way to a period of peace and stability unprecedented in European history" (Solana, 2003). Also, in 2003, the word "space" is nowhere mentioned in the EU security strategy. Whether Europe is indeed more under threat and a less safe place now than 20 years ago is beyond the scope of this book, but at least one could conclude that from an EU institutional perspective, the security of space has become an area of increased concern over the past two decades. Regardless, with extensive progress in satellite technology and the proliferation of its use, both civilian and military, it is no surprise that the role of space has grown in importance over the past two decades. While that is clearly visible in the area of civilian applications (transport navigation, telecommunications, even commercial space travel), it is also reflected in the evolution of the EU's security and defence policies and increased military use of space assets, for example the use of satellite imagery in support of EU military operations (Darnis, Veclani & Miranda, 2011). "Space is essential for the efficacy of modern armies, while its strategic aspect is undeniable", then Belgian Foreign and Defence minister Philippe Goffin told the European Space Conference in Brussels in January 2020. "But we remain opposed to the militarisation of space" (Tigner, 2020).

Similarly, French armed forces' space commander General Michel Friedling stressed there that "space should not become the new Wild West... but it is the security of our infrastructure in space that is at stake", as Europe's adversaries are creating weapon systems "that can attack or destroy our assets in space". "France seeks neither a belligerent position nor the militarisation of space, but we have to guard against hostile 'grey' actions or attacks that fall below the level of war", according to the space commander (Tigner, 2020). France has budgeted 5 billion euros for the period until 2025 to build a military infrastructure and ability to identify, detect and neutralise potential threats in space.

The notion of adversaries with bad intentions underlies much of the European Union's policy making in the area of space and security: while being a responsible player itself, Europe needs to prepare for hostile entities that could attack our space assets and interests. That rhetoric of increased and unpredictable threats is of course required to get sufficient support for an ever-expanding EU security and defence infrastructure. But with much of today's warfare inextricably linked with space systems – from communications and surveillance, to navigation for the guidance of missiles – the EU should be much clearer about the military potential of such systems, and what they understand to be the militarisation of space. That is a basic democratic obligation for a European Union that slowly keeps expanding its political and military powers.

NATO

And while the latest EU buzzword is "strategic autonomy", including in areas of security and defence, NATO is also pushing for a stronger space agenda on both sides of the Atlantic. In 2019, NATO declared space an operational domain, and in October 2020, it founded a new Space Centre housed within Allied Air Command (AIRCOM) at Ramstein Airbase in Germany, embedded within the Operations Directorate (Jennings, 2021). The Space Centre's mission is to help coordinate and support allied and NATO space activities and operations and also help protect allied space systems by sharing information about potential threats. Moreover, NATO's new "center for excellence" in military space is being set up at the Centre Spatial de Toulouse, also to be the headquarters of France's Military Space Command (Mackenzie, 2021). To further integrate the new domain, NATO has extended its Article 5 to apply to space, warning that attacks against allied space assets will risk triggering its mutual defence clause. In that regard, it is developing a "multidomain approach" to the defence and deterrence of attacks against its space assets. According to a military official quoted by *Jane's Defence Weekly*, such an approach to protecting space could mean shifting from defensive to offensive tactics (Tigner, 2021). NATO's "overarching space policy", released in January 2022, refers to partnerships with the EU, "on space and space-related aspects, as appropriate, and where it adds value to NATO's core tasks" (NATO, 2022).

An Emerging EU Space Policy

In 2001, then French President Jacques Chirac warned: "The United States spends six times more public money on the space sector than Europe. Failure to react would inevitably lead to our countries becoming first scientific and technological vassals, then industrial and economic vassals" (Slijper, 2012: 149). Clearly, Europe was doomed if it would not dramatically increase its space profile. While not always put so strongly, the need to catch up on bigger space powers has been a recurring theme throughout the past two decades. In close cooperation with the ESA, the European Commission and national capitals, Europe's military space agenda has developed to become a key element of the wider space policy since the beginning of this century.

Well aware of the sensitivity of any military aspects, the EU long preferred some ambiguity, with an emphasis on security rather than defence or even military. Since 2003, a series of documents shaped early EU space policy and created common ground on the use of space for security and defence purposes, culminating in the 2007 European Space Policy (Slijper, 2012: 149–154). The 2003 *White Paper* for example asserted:

> To be credible and effective, any CFSP and ESDP must be based on autonomous access to reliable global information so as to foster informed decision-making. Space technologies and infrastructures ensure access to knowledge, information and military capabilities on the ground that can only be available through the capacity to launch, develop and operate satellites providing global communications, positioning and observation systems.
>
> (European Commission, 2003: 12)

Also, it set out:

> space technology, infrastructure and services are an essential support to one of the most rapidly evolving EU policies—the Common Foreign and Security Policy (CFSP) including European Security and Defence Policy (ESDP). Most space systems are inherently capable of multiple use and the credibility of the above policies will be significantly strengthened by taking better advantage of space applications.
>
> (European Commission, 2003: 19)

Finally, it claimed:

> the European Space Policy (...) carries the promise of substantial economic, social and environmental benefits for the Union and its citizens. The policy will also bring new qualities to the Union's external actions, especially in defence, security, environment and development.
>
> (European Commission, 2003: 37)

In a 2005 letter to the European Parliament and the Council of Ministers, the Commission further elaborated its vision on military aspects of space policy:

> The Council of the EU has recognised that space assets could contribute both to making the EU more capable in the field of crisis management and to fighting other security threats. It has therefore approved the idea that identified and agreed upon ESDP requirements should be reflected in the global EU space policy and European space programme.
>
> (European Commission, 2005: 4)

These phrases are indicative for how Europe's early space policy was developed. When released in 2007, the European Space Policy – a joint product of the EC, and ESA – marked a clear departure from the past, when clear policy reference to the military dimensions of space had always been avoided (Slijper, 2012: 153). Under the "security and defence" header, the policy document mentions:

> The EU approach to crisis management emphasises the synergy between civilian and military actors. Space system needs for planning and conducting civilian and military Crisis Management Operations overlap. Many civilian programmes have a multiple use capacity and planned systems such as Galileo and GMES may have military users. The Member States in the Council have identified Europe's generic space system needs for military operations and stressed the necessary interoperability between civilian and military users. Military capability will continue within the remit of Member States. This should not prevent them from achieving the best level of capability, within limits acceptable to their national sovereignty and essential security interests. Sharing and pooling the resources of European civilian and military space programmes, drawing on multiple-use technology and common standards, would allow more cost-effective solutions. The economy and security of Europe and its citizens are increasingly dependent on space-based capabilities which must be protected against disruption. Within the framework of existing EU principles and institutional competencies, Europe will substantially improve coordination between its defence and civil space programmes, while retaining primary end-user responsibility for funding.
>
> (European Commission, 2007: 7)

While Galileo and Copernicus (GMES until 2008) were not yet operational, this reflection on the military dimensions of EU space policy has remained representative and has since become deeper embedded in ESDP/CSDP language.

In the Treaty of Lisbon, which entered into force in December 2009, article 189 refers to the importance of space for technical and scientific progress and joint initiatives in support of that, and to "coordinate the efforts needed for the exploration and exploitation of space" (European Union, 2017). Without explicit reference to military purposes, the general, open formulation in the treaty opens the door to further explore initiatives in that domain; initiatives which get the necessary push from Commission bureaucrats, industry representatives and politicians.

The integration of space into the ESDP has since then gradually evolved. As a 2009 DG Enterprise and Industry brochure put it:

> Space-based systems are making an increasingly large contribution to the security of Europe, and to the European Security and Defence Policy (ESDP) in particular. Europe faces constantly evolving threats to its security that are now more diverse, less visible and less predictable than in previous decades.
>
> (quoted in Slijper, 2012: 149)

On its website at that time, it further added ambition to that:

> Space systems are clearly strategic assets that demonstrate independence and the ability to assume global responsibilities. To maximise the benefits and opportunities that they can provide to Europe now and in the future, it is important to have an active, co-ordinated strategy and a comprehensive European Space Policy.
>
> (European Commission, 2009)

A few years later, in addition to the same phrase, the directorate expanded it with financial-material consequences to back up ambitions:

> Europe needs therefore to have access to the best affordable capabilities for autonomous political assessment, sound decision-making, prevention policies and the effective conduct of actions. Space assets provide a significant contribution to confronting these threats through global monitoring, communication and positioning capabilities.
>
> (European Commission, 2014)

The October 2016 *Space Strategy for Europe* again emphasises that space "is also of strategic importance for Europe. It reinforces Europe's role a stronger global player and is an asset for its security and defence":

> Growing threats are also emerging in space: from space debris to cyber threats or the impact of space weather. These changes make greater synergies between civil and defence aspects increasingly relevant. Europe must draw on its assets and use space capacities to meet the security and safety needs of the Member States and the EU.

Also it emphasises the need to reinforce synergies between civil and security space activities:

> Space services can strengthen the EU's and Member States' capacity to tackle growing security challenges and improve the monitoring and control of flows which have security implications. Most space technologies, infrastructure and services can serve both civilian and defence objectives. Although some space capabilities have to remain under exclusive national and/or military control, in a number of areas synergies between civilian and defence can reduce costs, increase resilience and improve efficiency. The EU needs to better exploit these synergies. This will be a key theme of the European defence action plan, which is expected to highlight space's crucial enabling role for civilian and defence capabilities.
>
> (European Commission, 2016)

Today, the Council of the European Union's *Why an EU space policy?* webpage actually puts security and defence rather central:

Helps create jobs and boost growth and investment in Europe; pushes back the boundaries of science and research; promotes and facilitates other policies in such areas as security and defence, industry and digital technology; plays a crucial role in the economic recovery after the COVID-19 crisis and in tackling global challenges such as climate change.

(European Council, 2021)

As set out by Oikonomou (2017):

The development of EU space policy and its two main programmes, Galileo and Copernicus, has necessitated a parallel process of legitimisation of this policy. Popularisation, defined as the simplification of a policy in order to be made accessible to the masses and accepted by them.

Although the military dimension has been less popularised, it can be argued that the dominant emphasis on its non-military purposes has contributed to its popularisation generally and has also contributed to obfuscating its military aspects, generally seen as the more controversial aspects of EU policy.

The ESA

Established in 1975, the ESA has been the embodiment of Europe's common space efforts. It is maybe best known for its involvement in the International Space Station (ISS), the so-called Spaceport in Kourou, French Guiana, and the Ariane family of launch vehicles. It has now 22 member states. The national bodies responsible for space in these countries sit on the ESA's governing Council, which also includes Canada, under a Cooperation Agreement. Not all EU states are members of the ESA and not all ESA Member States are members of the EU. Especially the latter was earlier identified as an "acute problem when it comes to security and defence matters", according to the European Commission (2012). Over time, it seems that both have acknowledged mutual synergies. While the Commission considers the ESA as a "technical manager" for its space programmes, the ESA "views the Commission's space budget as force multiplier" (De Selding, 2013).

While maintaining close ties with the EU through an *ESA/EC Framework Agreement*, aimed at avoiding the duplication of efforts, the ESA is an entirely independent organisation. The two organisations have worked together to develop the European Space Policy; since 2004, they met about annually within the so-called Joint Space Council. ESA programme boards each have their own field of activities, for example on communication satellites, satellite navigation, earth observation and launchers. Coherent with the security and defence perspective outlined in the EU space policy and increased cooperation with Brussels, the ESA gradually became more explicit regarding the security and defence dimensions of its work. While there used to be a rather clear distinction between the ESA's civil space agenda and the predominant national military space programmes, over the past two decades, the ESA has become involved in space projects with both civilian and military aspects.

Still, the clear stipulation in article 2 of the ESA Convention that "the purpose of the agency shall be to provide for and to promote, for exclusively peaceful purposes, cooperation among European states" required at least some further explanation as to how those exclusively peaceful purposes could be guaranteed with an expansion of its military footprint. ESA interprets "peaceful" use as "non-aggressive" (Praet, 2007), rather than "non-military". It has subsequently also been explained as follows: "ESA's security initiatives [...] must be provided for exclusively peaceful purposes, a provision which has been interpreted under international law as non-aggressive uses of outer space" (Domecq, 2017). This even narrower description appears to ignore the potential of aggressive uses enabled by assets in outer space.

This raises the obvious question of how non-aggressive is then defined for the obvious grey area covering what some would consider non-aggressive support by space assets to potentially offensive military action on the ground (or elsewhere). What if a country is bombed with intelligence heavily leaning on satellite imagery from European space assets, or a bomb is guided by Galileo navigational information? What are the boundaries and how will they be checked and enforced? "To determine how it might be involved in a European security programme without too obviously slipping the bounds of its charter" the ESA contracted various studies (De Selding, 2008). It is not clear though how these studies exactly shaped ESA's position, but a comment in early 2007 from its then-science director is at least illustrative: "Let's be realistic. Europe hasn't gone into space primarily to do science. The ability to put satellites in orbit is now an essential strategic and economic tool in telecommunications, navigation, security, etc." (Slijper, 2008: 36).

The EDA

The EDA, which became operational in 2004 through a Joint Action of the EU, has since become part of the ESDP infrastructure "driving, forward an ambitious agenda of cooperation" on a wide range of issues, including research and technology, armaments, military capabilities and industry (Linnenkamp, 2015: xv). As the Treaty of Lisbon envisions the progressive development of Europe's military capabilities, as well as taking forward Europe's space policy, and nurturing the military-industrial base, it is no incidence that the EDA plays a key role in fulfilling these agendas (Oikonomou, 2012). In 2013, Claude-France Arnould, then chief executive of the EDA, recalled that space "has become a vital element of any security or military action. [...] Space is now embedded in practically everything we do, in our everyday lives, but also in security operations, from intelligence gathering to flying a drone" (European Space Agency, 2013).

While the Commission together with the ESA have been leading most of the security- and military-related space initiatives – notably Galileo and Copernicus – EDA has taken up initiatives around specific military parts of space programmes, often together with either or both the ESA and the Commission. Space is a key priority for the EDA. Already in 2006, it noted: "the preparation and conduct of future EU led operations will require continued consideration of space related aspects, such as communication, and the detection and identification of potential threats in advance of an appropriate answer" (European Defence Agency, 2006).

In 2011, the EDA and the ESA signed an Administrative Arrangement for the establishment of cooperation between the two agencies, aiming to identify capability gaps that could be filled by space assets in support of relevant EU policies; to investigate whether capability requirement can be shared/supported by both agencies; to investigate synergies between existing and future EDA and ESA programmes; to coordinate research, technology and demonstration activities; to explore synergies and coordinate activities in support of industrial competitiveness. Other common areas of work identified included: Intelligence, surveillance, reconnaissance (ISR); civil-military synergies in earth observation; satellite communications (SatCom); space situational awareness (SSA); and critical space technologies for European non-dependence (European Defence Agency, 2011). At its November 2012 steering board meeting, EU Ministers of Defence tasked the EDA to "federate defence-related requirements, identifying civil-military synergies in preparation of demonstrators and by providing support to critical space technologies for European non-dependence" (European Defence Agency, 2012).

Putting words into practice has not always been easy. With a sense of understatement, the directors of the EDA and the ESA released a joint editorial in the 2017 special edition on space of EDA magazine *European Defence Matters*, where they argued:

Europe has been earmarking space and security as priorities for over a decade. Yet, it still hasn't fully lived up to its ambitions. There is now an unprecedented window of opportunity for addressing these shortcomings: a series of ambitious EU security related initiatives launched in 2016 can finally make space and security cooperation a tangible reality with positive effects on Europe's security, its economy and, perhaps even more important, on how citizens perceive Europe. (...) Synergies on dual-use amongst sectors make political, technological and budgetary sense. Such synergies have been called upon in most policy documents for over a decade, and again more recently in the European Commission's 2016 Space Strategy. But have we made real and genuine headway in fostering dual-use cooperation? Or have we perhaps not done enough to bring the two communities together and to build the confidence needed? We need ambitious policy statements, no doubt about that. But above all, we need action. Europe has never faced greater security challenges than today, be it at home, at its borders or in neighbouring countries. (...) For space and security to become a successful combination, ESA and EDA Member States need to further strengthen in partnerships and cooperation.

(Domecq, 2017)

Their apparent frustration about lack of progress and funding to further strengthen Europe's military space ambitions shows how developments in the area of CSDP often go slowly. Looking back over longer periods of time, tangible steps have however been taken.

Today, the EDA claims:

> By many accounts, there is a silent new Cold War taking place in space and the upper reaches of Earth's atmosphere as China, Russia, and the United States race against each other to enhance the usage of space for military tasks, while developing hypersonic weapon systems.
> (European Defence Agency, 2021b: 9)

The EDA also considers:

> Neither national nor multinational military operations are conceivable without the support of space-based systems. Space-based assets and applications are essential to navigation, communication, meteorological, geospatial and imagery services, early warning and ballistic missile interception. [...] Space is not only acknowledged as a potential theatre of operations, but also seen as an important strategic enabler in support of all other defence domains.
> (European Defence Agency, 2022)

The EDA's current space-related capability activities focus on SatCom, Space-Based Earth Observation (SBEO), Positioning, Navigation and Timing (PNT) and SSA. The EDA is also working closely with its Member States and partner organisations to coordinate the development of platforms and ground-space links in areas of "defensive planning and C2" [command and control] (European Defence Agency, 2021b: 9). In 2021, the EDA and the ESA completed the two-year study "Cyber Defence for Space" analysing cyber threats to ground and space infrastructure, recommending the creation of "cyber-operational centres services for cyber threat intelligence for space, utilising AI Technologies" (European Defence Agency, 2021a: 7).

A new area of work covered by the EDA is hypersonics, where research started in 2020 with a workshop on hypervelocity systems with experts from 12 Member States, plus Norway and Switzerland. The group concluded that countering hypersonic threats raises serious technical issues and that space-based tracking systems are probably the most suitable option for detecting hypersonic threats. Besides identifying a number of key research areas, the experts also noted that Europe needs new specialised facilities to test hypersonic technologies.

> While the USA and China each have three wind tunnel sites for Mach 6 velocity testing, Europe has none. Thus, a priority goal should be joint European test facilities among EDA countries. [...] The EDA is now preparing a study to investigate Europe's technological gaps, and will draw up technology roadmaps in 2022, with follow-up research projects focused

on missiles and munitions, electro-optical sensors, radar technologies, new materials, and guidance, navigation and control.

(European Defence Agency, 2021b: 9)

The Industry

Traditionally, there has been much emphasis on the need to foster Europe's space-industrial capabilities through EU space policy. As Margrethe Vestager, European Commissioner for digital policy, put it:

There are many more ways to get value for money [regarding the EU's space budget] by linking space to defence for dual-use purposes. [...] By that I do not mean militarising space but recognising that the civil space sector and defence share similar needs such as cyber defence or artificial intelligence. We should support those synergies.

(quoted in Tigner, 2020)

These interests are well-embodied in the European Commission's recently set up defence, industry and space policy department, DG DEFIS. "We cannot ignore the defence nature of what we do in space and we should not be naïve", according to Pierre Delsaux, its second in charge and veteran Commission official working in the area of military industry and space policy. "We need to defend our space assets to ensure that we are protected in space. And we need to be independent; we cannot remain dependent on the United States" (quoted in Tigner, 2020). Along such lines, the EU has stepped up its funding of space-related research and development projects, including in areas with security and defence purposes. These take currently place under the Horizon Europe (2021–2027) research framework programme, as well as through the European Defence Fund (EDF) and PESCO (Permanent Structured Cooperation) initiatives.

Until less than ten years ago, EU funding was off-limits for purely military purposes, while allowing for example research funding for "dual-use" security areas, covering both military and non-military applications. For example, research funding has been allocated under both the area of "space" and "security" under the 2007–2013 Seventh Framework Programme (FP7). Already as part of FP8, or Horizon 2020, it was slowly opened up for military research under the Preparatory Action on Defence Research (PADR), a precursor programme of the EDF, funding military research projects between 2017 and 2019 through EDA for a total of EUR 90 million. The European Industrial Development Programme (EDIDP) was another EDF precursor programme with a financial envelope of EUR 500 million for 2019–2020, the first ever EU grant programme co-financing the joint development of military products and technologies. From 2021, military technology funding has become fully part of the EU budget, with the EDF receiving close to EUR 8 billion for the 2021–2027 budget cycle.

One example of space-related research involving the EDA, research institutes and industry is called MIRACLE II (Micro-Satellite Clusters II), pointing to the "defensive potential of clustered imaging radar satellites" (European Defence Agency, 2021b: 9). The project was following up on work already initiated from 2004 by Kongsberg, Ericsson, Thales and then-Finmeccanica (now Leonardo) to "work on a system of micro-satellites that could achieve similar resolution and/or accuracy as larger satellites, but at lower cost" (MIRACLE, 2007). MIRACLE II, with Kongsberg and scientific institutes from Italy and Norway, funded by the EDA which with EUR 2.9 million, developed new concepts and technologies for clusters of imaging radar satellites. "MIRACLE II's architectural concepts for the satellites demonstrated how to maximise operational performance for spatial/temporal coverage and resolution, low vulnerability, and timeliness – all with direct application for Europe's militaries" (European Defence Agency, 2021b: 9).

Capabilities: Galileo, GOVSATCOM and Copernicus

The EU's largest military- and security-related space programmes are the Galileo satellite navigation system and Copernicus, the European earth observation programme.

In May 2021, the European Union Agency for the Space Programme (EUSPA), headquartered in Prague, was formed as EU agency managing the EU Space Programme and:

> To provide reliable, safe and secure space-related services, maximising their socio-economic benefits for European society and business. […] EUSPA is driving innovation-based growth in the European economy and contributing to the safety of EU citizens and the security of the Union and its Member States, while at the same time reinforcing the EU's strategic autonomy.
>
> (EUSPA, 2022)

EUSPA is responsible for operational management of the Galileo and Geostationary Navigation Overlay Service (EGNOS) satellite navigation programmes and for ensuring the continuous provision of their services. EUSPA also coordinates the EU's governmental satellite communications programme (GOVSATCOM).

A core task for EUSPA is the security of the EU Space Programme. This includes security accreditation of all components of the space programme, through the Security Accreditation Board. EUSPA is also responsible for the operational security of Galileo and EGNOS, which is provided through the Galileo Security Monitoring Centre, and for the provision and delivery of the Galileo Public Regulated Service (PRS) for governmental users (EUSPA, 2022).

While Galileo's first test satellite was already launched in 2005, it went live only in 2016 (early operational capability). Fully operational capability will consist of 24 active satellites and is now only expected in 2022. Galileo is operated by EUSPA, with two ground operations centres in Fucino, Italy and Oberpfaffenhofen, Germany.

The Galileo PRS is an encrypted navigation service for governmental authorised users and sensitive applications that require high continuity. Already at its inception, the desire for European military autonomy was an important reason for the EU to proceed with Galileo and thus it is clearly set to become a key asset for its military operations. An unnamed diplomat cited in 2008 on an EU news portal said, "everybody knows that there is no business case for Galileo. We only need a European system of our own, because at a militarily very critical moment we can't trust the GPS to be available" (Euractiv, 2008). With precision-guided missiles increasingly being guided by navigation satellites, Galileo could thus be set to become Europe's preferred system to guide bombs and missiles against any target perceived as a threat.

The GOVSATCOM programme aims at providing secure and cost-efficient communications capabilities for the EU and its Member States, including national security actors with a strong security dimension. Three main uses have been identified as crisis management; (border and maritime) surveillance; and key infrastructures, including EU space infrastructures such as Galileo and EGNOS. Similarly to Galileo's PRS, the GOVSATCOM users have to be authorised users to get access to services. The implementation of the GOVSATCOM component of the EU Space Programme started in 2021, under the new Space Programme Regulation, while the GOVSATCOM Preparatory Action, initiated by the European Parliament, started already in 2019.

In 2001, the European Council decided to establish what first became known as the Global Monitoring for Environment and Security (GMES) initiative, set up in 2005 by the European Commission and the ESA. It changed its name into Copernicus in 2012. According to the ESA:

> Copernicus is the most ambitious Earth observation programme to date. It provides accurate, timely and easily accessible information to improve the management of the environment, understand and mitigate the effects of climate change and ensure civil security. [...] ESA coordinates the delivery of data from upwards of 30 satellites. The EC, acting on behalf of the European Union, is responsible for the overall initiative, setting requirements and managing the services.
>
> (European Space Agency, 2022)

Logically, they could include the provision of satellite imagery for military and border control operations. "The EDA has gathered the satellite communication needs for European military actors involved in the conduct of common security and defence policy operations as well as for national military use", said an EDA spokesman in 2016 (Banks, 2016). Already in 2010, it was acknowledged that:

> Reflecting on current political dynamics, GMES stakeholders are now taking initiatives to strengthen the "S" in GMES by creating synergies between civilian and military actors. The 2008 EU Council Conclusions on GMES call on the Commission to foster the implementation of GMES security related services to support the related European Union policies. Border

surveillance, maritime surveillance and support to EU External Action have been identified as priority areas for action.

(EC/ESA, 2010)

Copernicus' costs between 1998 and 2020 are estimated at 6.7 billion euros with around 4.3 billion euros spent in the period 2014–2020 and shared between the EU (67%) and the ESA (33%). The 2016 Space Strategy highlights, "additional services will be considered to meet emerging needs in specific priority areas, including [...] security and defence to improve the EU's capacity to respond to evolving challenges related to border controls and maritime surveillance with Copernicus and Galileo/EGNOS"(European Commission, 2016).

Professor Anne Glover, the EU's former chief scientific adviser, believes that space systems, such as Galileo and Copernicus are increasingly becoming vital intelligence resources that will be used in planning and carrying out military missions. "It's already abundantly clear that the system will also be used for military operations and surveillance purposes", she said about Copernicus (Banks, 2016).

Beyond the Horizon: An EU Space Command?

Clearly, the past less than two decades have shown major developments in the security and defence thinking of the EU, and not less so in the area of space policy. The latest development has been the establishment of the Union Space Programme and the EUSPA, further expanding policy areas and activities outlined earlier in this chapter. If one thing is clear: the militarisation of space is no longer a taboo for the EU.

In his opening speech at the 14th EU Space Conference in January 2022, EU Commissioner Thierry Breton was especially bold in his statement regarding the EU's military ambitions in space, aiming at no less than a European Space Command. Besides referring to an upcoming *Space & Defence Strategy*, as part of the *Strategic Compass*, he elaborated extensively on how

> Beyond the traditional defence domains, we collectively face new threats in new strategic and contested areas. It is the case with cyber, it is also the case with space as space is crucial for our security. We should first expand the defence dimension in existing and upcoming EU infrastructures. Galileo is a clear demonstration that a common infrastructure under civilian control can meet defence and security needs, and that we can set up the right governance, based on trust. We should replicate this approach for the evolution of Copernicus, in the new secured connectivity initiative and of course in the STM [Space Traffic Management – FS]. Second, we should develop new infrastructures as dual-use by design, integrating the defence needs from the outset. [...] Finally, we should set up a new governance for our space programme to best reply to the threats. In addition to the established crisis management protocol, we could better organise joint situational awareness with

the participation of all the actors. Our aim on the mid- to long-run could be to establish a true European Space Command.

(Breton, 2022)

Conclusion

Space has become more accessible, including to smaller or technologically less advanced countries, making space more congested and increasing the risk of collision. It also increases strategic risk, with more states relying on space for their security interests and thus space assets becoming potential targets, invoking additional measures to secure these assets. Also, space has gradually become a key enabler in warfare. Positioning, navigation and timing have become core functions for weapon targeting and delivery and will continue to be further developed. Advances in observation capabilities, including through increased uses of rapidly developing sensor and artificial intelligence applications, will enable an ever-growing information potential that can be relayed by more powerful communication satellites. The classical function of space to support warfare is set to expand. Thus, the space domain has both become indispensable in supporting military operations and is itself developing into a theatre for military operations.

In less than 20 years time, space has become a cornerstone of the European Union's security and defence policies. With that it has become clearer what role space plays in that area, though many questions about its ambiguity have remained: when does defensive militarisation become offensive, and when is peaceful military use contributing to warfare? The ambiguity towards the militarisation of space is clearly reflected by the support of EU nations of UN initiatives to prevent an arms race in space, while at the same time their militaries have become increasingly reliant on space infrastructure to support military missions. It seems logical that no state would want military escalation in space, and yet we witness states that have demonstrated anti-satellite capabilities using ground-based weapons systems. We also see that jamming of SatCom traffic is on the rise. Further escalation can be expected without common efforts by the international community to prevent that.

While it is logical for the EU to plan and prepare against attacks – which the EU calls "defensive militarisation of space" – it is also clear that it will not stop an arms race in space. For that, Europe should use its diplomatic clout more and step up its role in multilateral initiatives, preferably under UN banner. As set out during the 2020 UN General Assembly First Committee sessions by civil society:

> support for the Prevention of an Arms Race in Outer Space – PAROS – remains strong. But the divides over how to implement this objective – whether through legal restrictions, political commitments or normative understandings of responsible behaviour – remain equally strong. These are not mutually exclusive options. None can progress without efforts to enhance trust and transparency.

(West, 2020b)

So, rather than focusing on military responses which may lead to counter-responses and a further build-up of tensions, such practical and feasible steps in the area of diplomacy are:

> Rooted in existing norms of behaviour that could be taken in the short term to enhance security in outer space. Such steps would increase the transparency of military space activities and help to build the trust needed to support long-term arms control measures.
>
> (West, 2020a)

Bibliography

Banks, Martin. 2016. "Mixed Reviews on EU Plan to Use Commercial Space Assets for Military." *Defense News*, August 3, 2016. Available from: https://www.defensenews.com/air/2016/08/03/mixed-reviews-on-eu-plan-to-use-commercial-space-assets-for-military/

BBC. 2008. "US Spy Satellite Plan 'a Cover'". *BBC*, February 17, 2008. Available from: http://news.bbc.co.uk/2/hi/americas/7248995.stm

Breton, Thierry. 2022. "Speech by Commissioner Thierry Breton at the 14th EU Space Conference." January 25, 2022. Available from: https://ec.europa.eu/commission/presscorner/detail/es/speech_22_561

Darnis, Jean Pierre, Anna Veclani, and Valérie Miranda. 2011. "Space and Security: The Use of Space in the Context of the CSDP". *European Parliament*, October 30, 2011. Available from: https://www.europarl.europa.eu/thinktank/en/document/IPOL-SEDE_ET(2011)433834

De Selding, Peter. 2008. "France to Keep Pushing for New EU Space Policy." *Defense News*, July 14, 2008.

De Selding, Peter. 2013. "Resolution Underscores Complications in ESA-EU Partnership." *Space News*, February 19, 2013.

Domecq, Jorge, and Johann-Dietrich Wörner. 2017. "Space and Security: Crucial Synergies For European Citizens." *European Defence Matters*, Issue 13, 2017. Available from https://eda.europa.eu/webzine/issue13/cover-story/space-and-security

Euractiv. 2008. *Galileo*. Updated version of April 23, 2008. Available from: http://web.archive.org/web/20120107224717/http://www.euractiv.com/science/galileo/article-117496

European Commission. 2003. "White Paper—Space: A New European Frontier for an Expanding Union." COM(2003)673, November 11, 2003.

European Commission. 2005. "Communication from the Commission to the Council and the European Parliament: European Space Policy—Preliminary Elements." COM(2005)208 final, May 23, 2005.

European Commission. 2007. European Space Policy. COM(2007) 212 final, April 26, 2007.

European Commission. 2009. "Bringing Space Down to Earth." D*G Enterprise and Industry website*, May 2009. Retrieved January 29, 2022 from http://web.archive.org/web/20090527093006/http://ec.europa.eu/enterprise/policies/space/index_en.htm#top

European Commission. 2012. "Establishing Appropriate Relations Between the EU and the European Space Agency". COM(2012) 671. November 14, 2012.

European Commission. 2014. "Space and Security". DG Enterprise and Industry, updated April 3, 2014. Retrieved January 29, 2022 from http://web.archive.org/web/20140406110400/http://ec.europa.eu/enterprise/policies/space/policy/space-security/index_en.htm

European Commission. 2016. "Space Strategy for Europe." COM(2016) 705 final, October 26, 2016. Available from: http://ec.europa.eu/docsroom/documents/19442.

European Commission, European Space Agency. 2010. "Space and Security." *EC/ESA Joint Secretariat Paper*, February 2010.

European Council. 2021. "EU Space Policy." Available from: https://www.consilium. europa.eu/en/policies/eu-space-programme/# [accessed July 22, 2021].

European Defence Agency. 2006. "An Initial Long-Term Vision for European Defence Capability and Capacity Needs." October 3, 2006.

European Defence Agency. 2011. "EDA & Space." *EDA Fact Sheet*, June 17, 2011. Available from: http://www.eda.europa.eu/docs/documents/factsheet_-Defence_space_final. pdf [accessed January 29, 2022].

European Defence Agency. 2012. "Interaction Between Defence and Wider EU Policies." November 19, 2012. Available from: http://www.eda.europa.eu/info-hub/news/2012/11/19/ interaction-between-defence-and-wider-eu-policies [accessed January 29, 2022]

European Defence Agency. 2021a. "Artificial Intelligence. How EDA Contributes." *European Defence Matters*, Issue 22, 2021. Available from: https://eda.europa.eu/docs/ default-source/eda-magazine/edm22singleweb.pdf

European Defence Agency. 2021b. "Space and Hypersonic Weapon Systems." *European Defence Matters*, Issue 22, 2021. Available from: https://eda.europa.eu/docs/default-source/eda-magazine/edm22singleweb.pdf

European Defence Agency. 2022. *Space* webpage at Available from: https://eda.europa.eu/ what-we-do/capability-development/space [accessed January 29, 2022].

European External Action Service (EAAS). 2021. "A Strategic Compass to Make Europe a Security Provider". *Foreword by HR/VP Josep Borrell, European Commission*, HR(2021) 174, November 9, 2021.

European Space Agency. 2013. *Claude-France Arnould, Chief Executive of the European Defence Agency: 'Space Is Now Embedded in Practically Everything We Do.'* July 22, 2013. Available from: http://www.esa.int/About_Us/DG_s_News_and_Views/Claude-France_Arnould_Chief_Executive_of_the_European_Defence_Agency_Space_is_now_embedded_in_practically_everything_we_do/

European Space Agency. 2022. "Europe's Copernicus Programme." Available from: https:// www.esa.int/Applications/Observing_the_Earth/Copernicus/Europe_s_Copernicus_ programme

European Union. 2017. "Treaty of Lisbon Amending the Treaty on European Union and the Treaty establishing the European Community, Signed at Lisbon, December 13, 2007." Available from: https://eur-lex.europa.eu/eli/treaty/lis/sign

EUSPA. 2022. Available from: https://www.euspa.europa.eu/about/about-euspa

Gohd, Chelsea. 2021. "Russian anti-Satellite Missile Test Was the First of its Kind". *Space.com*, November 17, 2021. Available from: https://www.space.com/russia-anti-satellite-missile-test-first-of-its-kind

Jennings, Gareth. 2021. "NATO Selects Six 'Space Pitch' Qualifiers". *Jane's Defence Weekly*, June 9, 2021.

Linnenkamp, Hilmar. 2015. "Foreword". In *The European Defence Agency – Arming Europe*, edited by Nikolaos Karampekios and Iraklis Oikonomou, xv–xviii. Routledge.

Mackenzie, Christina. 2021. "NATO Names Location for New Military Space Center". *Defense News*, February 5, 2021. Available from: https://www.defensenews.com/ space/2021/02/05/nato-names-location-for-new-military-space-center/.

MIRACLE. 2007. "Research and Technology Project "Micro-*Satellite* Cluster Technology". Available from: https://eda.europa.eu/docs/documents/MIRACLE_Project_-_Final_ Report.pdf

NATO. 2022. "NATO's overarching Space Policy". *NATO*, January 17, 2022. Available from: https://www.nato.int/cps/en/natohq/official_texts_190862.htm.

Oikonomou, Iraklis. 2012. "The European Defence Agency and EU Military Space Policy: Whose space odyssey?" *Space Policy*, Volume 28, February 2012.

Oikonomou, Iraklis. 2017. "'All u Need Is Space': Popularizing EU Space Policy." *Space Policy*, Volume 41, August 2017.

Praet, Michel. 2007. "The European Union and ESA. Issues Linked to Security/Defence." *ESA Presentation at NATO Defence College*, January 31, 2007.

Slijper, Frank. 2008. "From Venus to Mars. The European Union's Steps Towards the Militarization of Space." *TNI/Campagne Tegen Wapenhandel*, November 2008. Available from: https://www.tni.org/files/download/venustomars.pdf.

Slijper, Frank. 2012. 'Space: The High Ground of the European Union's Emerging Military Policies'. In *The Marketing of War in the Age of Neo-Militarism*, edited by Kostas Gouliamos and Christos Kassimeris, 145–169. Routledge.

Solana, Javier. 2003. "A Secure Europe in a Better World". December 8, 2003. Available from: https://data.consilium.europa.eu/doc/document/ST-15895-2003-INIT/en/pdf.

Tigner, Brooks. 2020. "EU wants Only Defensive Militarisation of Space". *Jane's Defence Weekly*, January 29, 2020.

Tigner, Brooks. 2021. "Alliance Chiefs Use Article 5 to Protect Space Domain". *Jane's Defence Weekly*, June 23, 2021.

West, Jessica. 2020a. "Did Russia Test A Weapon In Space?". *Ploughshares Spotlight*, July 2020.

West, Jessica. 2020b. "Joint Statement on Outer Space". *Project Ploughshares*, October 13, 2020. Available from: https://reachingcriticalwill.org/images/documents/Disarmament-fora/1com/1com20/statements/13Oct_space.pdf

3 The Militarisation of Outer Space

A European Perspective

Pascal Legai

On 7 September 2018, Madame Florence Parly, French Minister for the Armed Forces, announced the creation of the French Space Command under military governance (JORF, 2019), similar to the US model, in order to implement the national space strategy. Likewise, North Atlantic Treaty Organization (NATO) decided to declare space as a full-fledged operational domain at the same level with land, sea, air and cyberspace.[1] These essential orientations from European nations (the UK and Germany also created a national military space command) and NATO are the result of the growing dominance from the major space powers, mainly Russia and China, particularly in spy acts in space and anti-satellites (ASAT) capabilities. The militarisation of space is underway. The European Union (EU) anticipated this worrying evolution. In 2016, the EU started fostering space developments to significantly contribute to the effectiveness of the EU external action, including security and defence purposes. The increasing, multifaceted, transnational threats described in the EU Global Strategy for Foreign and Security policy of June 2016 (EUGS, 2016, 19–28) raise the question of the policies to be implemented to tackle them, including space-related activities. This orientation opened an active debate among the European space actors identified in article 189 of the Lisbon Treaty: the EU, Member States and European Space Agency (ESA) within their respective remits, highlighting in particular the growing role of the European Commission with the Space Strategy for Europe in October 2016 (European Commission, 2016a) and the European Defence Action Plan in November 2016 (European Commission, 2016b), and the Regulation establishing the space programme of the Union adopted by the European Parliament and the Council in early 2021 (European Parliament & Council, 2021). The current evolution of a stronger Union in Europe questions the historic national prerogatives of sovereign responsibility in the security and defence fields that the governmental space policies underpin (Gnesotto, 2009, 29–30).

Thus, the European space policy, based on the idea that "Security from space" requires first Security in Space",[2] sets up a unique space programme in its Space Regulation encompassing a comprehensive set of dual-use components: Galileo, Copernicus, SSA/SST, Govsatcom and launchers (European Parliament, Council, 2021, preliminary part). It implies the preservation of a world-class research and innovation level, a state-of-the-art industrial and technological base as part

DOI: 10.4324/9781003230670-5

of the crucial strategic autonomy (Crespi, Caravella, Menghini, Salvatori, 2021, 348–354). In this regard, the space budget for the period 2021–2027 also remains at a credible level (14.9 billion euros) (Council, 2021, 1). We can also add that the recovery fund (Next Generation EU, NGEU) of 750 billion euros (General Secretariat of the Council, 2020, 1–6) adopted by the European Council on 21 July 2020 will be partially used at national level for space activities. Furthermore, the EU has not managed until now to promote a peaceful use of outer space through an internationally recognised legal framework. Within this context, ESA adds its fundamental longstanding experience, despite its Convention excluding defence and military activities, in the design, development and procurement of space systems. This comes in addition to its innovation know-how for downstream applications dealing with huge amounts of space and non-space data. The new ESA Director General, Josef Aschbacher, in office since 1st March 2021, gave a strong orientation towards "Safety & Security" in the ESA Agenda 2025 (ESA, 2021, 9–10).

The circumterrestrial space beyond the aerobic layer has become a new place of confrontation, of power, of hostile or aggressive action. This is a potential battle field which can directly affect the daily life of the citizens of the world, the planetary economic development whose dependence on space is considered vital today, as well as science and exploration activities. The concept of the militarisation of space therefore takes on its full meaning when the confrontation that takes place affects the means dedicated to the defence of certain states. Control of the space environment therefore constitutes, in a way that is perfectly accepted by all, an issue of power, sovereignty and strategic autonomy. The EU is well aware of this (European Parliament, 2020, 31–34).

This chapter means to elaborate on these different intertwined and sometimes contradictory issues to define a European future between submission and independence. It considers first the current European legal debate on the issue of militarisation evolving towards weaponisation of Space, then the EU initiative to promote a non-binding international code of conduct in outer space and, finally, the current political position of Europe towards these issues will be considered in turn.

A Legal Approach in Europe to Militarisation and Weaponisation of Space

In Europe, that is the EU and its Member States, an active debate started around a decade ago as regards the progressive issue of militarisation of outer Space (Duffort, 2020) leading to the potential placement of weapons in space and a potential new arms race. First and foremost, this debate focused on the key element of the distinction between the militarisation and weaponisation of outer space, like in other major space-faring nations. Then, the debate moved to the related European political, R&D, innovation, industrial and operational orientations.

On the one hand, it is generally admitted, in particular in Europe, that the militarisation of outer space is the placement of satellites in orbit to support military activities on the ground (Tripathi, 2013, 193–194). Today, militaries all over the world rely on satellites for command and control, communication, intelligence, early warning, weather forecast and navigation/positioning/timing (PNT) from the Global Navigation Satellite Systems (GNSS). Therefore, "peaceful uses" of outer space include military uses, even those which are not at all peaceful—such as using satellites for direct bombing raids or to orchestrate a "prompt global strike" capability (Reaching Critical Will, 2014), which is "the ability to control any situation or defeat any adversary across the range of military actions" (US Department of Defense, 2004). On the other hand, the notion of space weaponisation refers to weapons being placed in space (Tripathi, 2013, 193–194) that have a destructive capacity to target locations or objects on Earth and in space, and those on Earth capable of targeting space assets, weapons that transit in outer space in order to reach their targets, such as hypersonic technology vehicles or ballistic missiles, can also be considered to be part of the weaponisation of space.[3]

Based on these general definitions, an important question is then a more precise characterisation, or typology, of a weapon in space (Perez, 2019, 6–7). The problem with the latter is that there is no globally agreed definition of what space weapons are. One of the dangers in outer Space is that almost anything can be used as a weapon. It includes weapons that can attack space systems in orbit (i.e. ASAT weapons),[4] attack targets on Earth from space or missiles travelling through space. We can add conventional kinetic, nuclear or bacteriological weapons and the control of a satellite by its motors or of debris deviated with laser techniques against another space object. Cyberattacks also make it possible to jam, alter and listen to communications. Blinding lasers can also put space assets permanently out of use. We can add airborne lasers (e.g. attached to Boeing 747-400 aircraft) or the Tactical High Energy Laser (THEL), which projects enough energy at flying rocket warheads to cause them to detonate.[5] The wilful generation of a debris wall can also be a type of weapon by modifying the environment, leading to avoidance manoeuvres.[6] Hence, the inability to define space weapons other than a list of possibilities is the main barrier to a treaty that prevents them.

Furthermore, space weaponisation is not a new phenomenon (Rodhan, 2018). However, a large number of technological developments over the past few decades have led to a drastic acceleration in the destructive potential of space warfare. The Prompt Global Strike program is a project within which the United States started developing hypersonic glide vehicles in secret in the mid-2000s. Such hypersonic glide vehicles are different from conventional ballistic missiles in three ways. First, they have a longer range and can travel over half of the Earth's circumference. Second, they can approach their target from a direction opposite to the expected trajectory of a typical ballistic missile and do so on a low-altitude gliding trajectory within the atmosphere. Third, they can be extremely precise, with terminal guidance systems enabling them to strike with an accuracy of a few meters. These characteristics make such vehicles nearly impossible to detect. Such missiles could effectively decimate a country's nuclear and military arsenals in a few tens of minutes, using low-yield nuclear weapons or even conventional explosives. Thus, the precision of hypersonic weapons eradicates the nuclear deterrent (Aarten, 2020).

Beyond the difficult definition of a space weapon, the question arises of the characterisation of an act of aggression. It is by definition the use of a weapon with the difficulty of demonstrating intentionality (Legai, 2021, 194). One must distinguish attacks from the ground to space (Anti-Satellites, ASAT),[7] or from an aircraft, from attacks only in space, or on Earth against the ground segments of space systems. Is scattered debris the result of an intentional act or simply the consequences of an accident? Is the collision between two objects in orbit accidental or intentional? What about an act of bringing together satellites ("browsing")? Do these acts amount to an act of war? The definition of an act of aggression is also not clear. The whole difficulty therefore lies in the definition of "spatial conflictuality" (Legai, 2021, 194).

Furthermore, the notion of territory does not exist in space, because the 1967 Treaty establishes the principle of non-appropriation of outer space. Thus, no state has the right to act to protect its spacecraft in outer space by means of military space operations of active defence, in contradiction with the orientations of certain states such as the United States or France in particular (Parly, 2019, 27–29).

The uncertainties of this legal debate have an impact on the related European political, R&D, innovation, industrial and operational orientations to anticipate tensions or crises in space. The most significant example is that of the development of Space Surveillance and Tracking (SST) capabilities which give a major strategic advantage to powers capable of precisely knowing the spatial situation of objects and debris in orbit and the effects of space weather in particular. This ability allows you not only to protect your own systems on which so many downstream applications depend but also to know the adverse capabilities present in outer space. Some EU Member States with space observation capabilities have committed to providing SST services[8] to authorised EU users. These services cover the provision of anti-collision information between space objects, fragmentation information following a collision in space and warning of re-entry of space objects into the atmosphere. This support is based on the existing capacities of certain Member States. A first consortium of five contributing nations has thus been set up (France, Germany, Italy, Spain, UK), supplemented since by three other nations (Poland, Portugal, Romania).

However, this approach shows its limits. The surveillance of space is a subject considered to be very strategic by the Member States that have devoted significant resources to this area. It gives them credibility vis-à-vis other major space nations. The pooling of these capacities for the benefit of the EU and its Member States is therefore done in a measured and prudent manner without having shown, for the moment, any real added value (Legai, 2020). To strengthen national autonomy in space surveillance, some European states have decided to create a space command, under military governance, such as France (JORF, 2019), the United Kingdom (RAF, 2021) or Germany (German MoD, 2021), whose main vocation is to implement a space strategy of protection of in-orbit capabilities.

In summary, if there is no territoriality within the meaning of the 1967 Outer Space Treaty (Legai, 2020), nor a perfectly clear concept of a weapon and a characterisation of an act of aggression in outer space, it is difficult for states to justify defending themselves in this context. Indeed, the concepts of weapon, aggression and territory are the constituents of the invocation of international self-defence within the meaning of the Charter of the United Nations (art. 51). In theory, if we prohibit acts of aggression, we de facto prohibit the use of preventive self-defence, and therefore, we prevent any risk of security escalation. If outer Space is *a priori* exempt from any sovereignty, international law still has a vocation to regulate it. In the absence of a treaty on armed conflict in space, it would seem that International Humanitarian Law (IHL), or the law of war, customary law resulting from the observation of practice, would be the most suitable to frame conflicts in space or of space origin with consequences on Earth. The need for international space regulation is intrinsically linked to the apprehension of an armed conflict in space. For the time being, a use of armed force or an armed aggression in outer states has not been observed, within the meaning of jus ad bellum, that is referring to the conditions under which states may resort to war or to the use of armed force in general. In this uncertain legal context, some European Member States (France, Germany, UK) have decided to acquire national space surveillance means mainly relying on their own R&D considered critical knowledge in order to put in place sovereign operational capabilities for enhanced space defence ensuring their strategic spatial autonomy.

A Major Initiative of the EU

Therefore, not only to avoid a militarisation of space and its drift towards weaponisation but also to manage the ever-increasing space traffic (Space Traffic Management, STM), space law has endeavoured to evolve since the 1967 Treaty and the Space Conventions[9] under the umbrella of Committee on the Peaceful Uses of Outer Space (COPUOS).[10] However, space law is struggling to make progress in the face of the reluctance of certain states, which see it as a limitation of their freedom of action, or even of their strategy of domination. Thus, in 2014, the initiative of the EU to establish an international code of conduct for outer space activities was unsuccessful (EU, 2014).

Indeed, the EU and its Member States have submitted to the international community a proposal for an international code of conduct for human activities in space (EU, 2013; EU, 2014) in order to adapt the legal framework to the new challenges of space but fundamentally to preserve this environment by limiting any use detrimental to the well-being of humanity. This proposed text, without much effect to date, adds three principles to space law existing since the 1967 Outer Space Treaty regulating human activities in outer space for peaceful purposes:

- improving the security of space operations and reducing space debris,
- creating the right to individual and collective self-defence,
- establishing a climate of trust and cooperation through the transparency of the respective activities of all actors. The legal path can also be a means of regulation and influence, if it is accepted by all.

Thus, in 2008, the EU initiated a procedure to develop an International Code of Conduct for Outer Space Activities (ICoC). The code was not intended to function as a legally binding treaty but is meant to consist of a set of principles and guidelines agreed to on a voluntary basis amongst states. It is not intended to have any formal enforcement mechanisms. Once agreed upon, the EU has stated (Johnson, 2014) that it expects the ICoC to be applicable to all outer space activities conducted by states, corporations, universities etc. and present the basic rules for both civil and security space activities. The code is intended to address both safety and sustainability of space environment and the stability and security in outer space, safeguarding all countries' inalienable right to use space for peaceful purposes.

Since it is aimed at both safety and security of outer space activities, the EU stated that existing internationals such as the Conference on Disarmament (CD) and the United Nations COPUOS are not appropriate for the ICoC. By discussing the ICoC outside the CD and COPUOS, it also includes UN Member States which are not members of these bodies. The EU has stated that it believes the non-legally binding and overarching nature of the ICoC means it does not contradict any ongoing discussions on for example Prevention of an Arms Race in Outer Space (PAROS).

The main goal of this European initiative is to "find an agreement on a text that is acceptable to all interested States and that thus could produce effective security benefits in a relatively short term" (Johnson, 2014). The support from the international community is the following: Australia, Canada and Japan have already endorsed the ICoC, while others have been less positive. Countries such as Brazil, Russia, India and China have expressed disappointment about not having been sufficiently consulted in its development. Together with other space emerging countries, they also raised concerns that the ICoC could be a way to limit their future capacities for further outer space activities. India's main issue with the code is that it is not legally binding, with enforcement, verification and a penalty mechanism (Listner, 2012). The United States, the leading country in space development, endorsed the ICoC after having had a national debate where some concerns were raised that the ICoC could lead to the mistaken belief that it could constrain missile defences or ASAT weapons (Reaching Critical Will, 2014). Other criticisms raised have been that it replicated already existing domestic policies from some of the EU Member States or in bilateral and multilateral transparency and confidence-building measures (TCBMs) (Reaching Critical Will, 2014). This criticism is based on the fact that the joint ICoC can be seen as interference in the domestic policy-making of nations that are already developing outer space policies on their own initiative.

However, the Code has been praised since it can be applied to all types of outer space activities as mentioned in Section 1.2 and, therefore, not only is a tool for environmental protection but also includes arms control aspects. Secondly, the ICoC also addresses military activities in outer space directly through Section 4.2, where the subscribing states commit to refraining from any action:

> Intends to bring about, directly or indirectly, damage, or destruction, of outer space objects unless such action is conducted to reduce the creation of outer space debris and/or is justified by the inherent right of individual

or collective self-defense in accordance with the United Nations Charter or imperative safety considerations.

(EEAS, 2014)

This means that the Code limits the testing and use of space-based and ground-based ASAT Weapons. The open-ended consultations in Kiev in May 2013 were the first multilateral meeting held on the draft ICoC. The meeting aimed at getting different states on the same level of information and knowledge. At the end of the two-day consultation, the EU announced that the next step would be to review all the participants' concerns and opinions in order to incorporate as many views as possible into the Code. The second open-ended consultations took place in Bangkok in November 2013. The Bangkok meeting focused on the actual content and wording of the proposed text, including the Preamble, Purposes, Scope and General Principles. A new revision of the draft based on the Bangkok consultations was realised on 31 March, 2014. This draft was the basis for the third consultations that took place in Luxembourg on 27–28 May, 2014.

In terms of international recognition, the Group of Governmental Experts (GGE),[11] whose final goal is to deliver a consensus report[12] that outlines conclusions and recommendations on TCBMs (Robinson, 2010, 10–11) for space security and sustainability, built its work on previous and on-going space security initiatives, in particular the EU's International Code of Conduct initiative. The GGE also based its work on the previous GGE from 1991 to 1993, the 1967 Outer Space Treaty, the UN COPUOS's Long-term sustainability of outer space activities (LTSSA) Working Group and already established bilateral TCBMs.

Despite a widespread international recognition that the existing regulatory framework is insufficient to meet current and future challenges facing the outer space domain, the development of an overarching normative regime comes up against strong resistance from a number of states that see it as a potential limitation of their activities as shown by the failure in 2014 of the EU initiative on a code of conduct in outer space (Beard, 2016). Progress has been made for sustainability and safety. Measures of transparency and confidence-building constitute a complementary approach, mainly based on a voluntary basis, thanks to technology transfer, capacity building and growing cooperation. International governance is essential, requiring a real will of the international community. An international space agreement, within the spirit of the Paris Conference in 2015 on Climate Change, is likely to support the long-term sustainability of outer Space as a secure and peaceful space environment. As an example, the COPUOS released guidelines for the long-term sustainability of outer space activities on 22 June, 2019 (UNOOSA, 2019). In September 2018, Stéphane Israël, Arianespace CEO, stated: "space should not be a new Wild West. This conquest imposes rules for sustainable Space".[13] To set an appropriate global context to progressively move towards comprehensive control to prevent a stronger militarisation and weaponisation of space, an internationally recognised regulatory legal framework turns out to be a decisive step. This necessary evolution would ensure the proper governance and equitable access to and sustainability of this environment. Debris removal and the obligation to provide for the return to Earth of any object before it is launched are current orientations. In this regard, ESA signed a contract in November 2020 with the Swiss start-up ClearSpace to start cleaning space from ESA Low Earth Orbit debris (ESA, 2021), the first such initiative in the world. Furthermore, to manage the fast-growing space traffic, space law has been evolving since the UN Treaty of 1967. As such, we can cite the space Agreements under the aegis of the COPUOS, including the Agreement on the Rescue of Astronauts and the Return of Objects Launched of 1969 (UNOOSA, resolution 2345 (XXII), 1967) and the Convention on the Registration of Objects Launched into Outer Space of 1975 (UNOOSA, resolution 3235 (XXIX), 1975) to establish a common space picture, that is to share the same knowledge on all objects in space.

European Strategic and Political Strand

The EU is not building its space policy on specific military or defence capacities. Its approach is resolutely civil for the benefit of the greatest number, for economic development, for growth, for employment and for the internal market as clearly expressed in the Space Strategy for Europe (European Commission, 2016a). Furthermore, in terms of security and defence, the Union shows a high level of ambition in order to implement the associated policies according to its Global Strategy which cannot be realised without essential mastery of the space environment (EUGS, 2016).

Space in Europe is now seen as an essential means of implementing the EU Global Strategy. This strategy became quickly the subject of an implementation plan in order to materialise the announced ambition (Council, 2016). This plan defines three strategic priorities to face threats and risks with the appropriate capacities and structures: reactivity to crises and external conflicts, contribution to the development of the capacities of partners and protection of the Union and its citizens. The Common Security and Defence Policy (CSDP) is taken into account in its civilian and military aspects. Within the framework of the European Defense Agency (EDA), a capacity development plan aims to fill the gaps through increased cooperation of the Member States, in particular the structures of EU intelligence, planning and conduct of operations (EDA, 2018). The capability priorities identified include Intelligence, Surveillance and Reconnaissance (ISR), drones, satellite communications, autonomous access to space and permanent observation of the Earth. The major powers like the United States, Russia, China and India have decided to equip themselves with the complete panoply of the immense capacities that the space domain offers. Europe, that is to say the EU and its Member States, with the significant contribution of ESA through its Agenda 2025 including a Safety and Security priority (ESA, 2021, 9–10), has also defined a coherent and comprehensive space policy to acquire the space resources necessary to cover its security and defence needs in its space Regulation (European Parliament & Council, 2021). Indeed, a Member State alone could not today acquire all necessary capacities and, above all, keep them at the state-of-the-art, unless it defined space as a major priority to the detriment of other sectors.

Furthermore, this space policy covers all stages from design, development, deployment and procurement to an innovative and competitive European industry supported by the Union. It defined the vision of the strategic issues for Europe to which space can contribute, compared its approach with its Member States as to the idea of a space that would serve dual needs, civil and military, without ever evoking a space which would be specifically military as we can see in the Defence Action Plan (European Commission, 2016b). Through the European space policy linked to the Common Security and Defense Policy (CSDP), it is also the means for the Union to assess the solidarity and the will of its Member States. In addition, it is a way to validate models of sharing activities and capacities between community and national actors, between the public and private sectors, and ultimately to assert its credibility on the international scene, as space today makes a decisive contribution to meeting security and defence needs.

In this context, the European Commission, whose fundamental vocation is to develop capacities and free services for the general public, is now becoming an actor in the areas of security and defence by having created a General Directorate for "Defense Industry and Space" (European Commission, 2021a). An essential step will be to consider setting up procedures and systems capable of handling and managing sensitive or classified data, products and services, and integrating the notion of confidentiality for authorised users. This is a major but inevitable development that undoubtedly constitutes a decisive step regarding a military and defence vocation for the EU. Indeed, the European space policy led by the European Commission aims at putting in place a complete range of space assets. The European approach is above all civilian, but each of the components has dual capabilities, civil and military, such as access to space (launchers), Galileo, Copernicus, SST and Govsatcom. Moreover, the Commission plays a general role of coordinator and federator as defined in the space Regulation (European Parliament & Council, 2021), seeking to avoid duplication by encouraging synergies, with the support of the European Defence Agency (EDA) for defence aspects. This approach can only succeed through the real will of the Member States to contribute to it. The strategic issues structuring this space policy are above all the autonomy of situation assessment and action contributing to the credibility, reliability, responsiveness and relevance of the EU on the international scene. This essential European autonomy is based upstream on maintaining a world-class level in R&D, advanced technologies and innovation, supported by a solid defence industrial base, the essential fabric of Small and Medium Enterprises (SMEs) and start-ups, business incubators and accelerators which are so fundamental to creating the added-value derived from space technologies. Autonomy also means controlling the entire chain from the sensor in orbit, transmission to the ground, processing and development of products and services, from secure and protected dissemination to the end user. In addition, the real political will is reflected in the allocated resources. In its budget for the period 2021–2027 (Multi-annual Financial Framework, 2021–2027) (European Commission, 2021b), the EU gives prominence to space with a budget of over 14.9 billion euros with a clear priority for Galileo (9.1 billion euros) and Copernicus

(5.42 billion euros), to the detriment of GovSatcom and SSA/SST (442 million euros for both) (EARSC, 2020).

Conclusion

In conclusion, the EU's spatial vision, supported by its Member States and ESA, is therefore not approached through the prism of a "military" space and related strategic issues, and not at all with an idea of arsenalisation of space, but as the establishment of structures and a range of complementary instruments, through increased cooperation between nations to meet collectively identified security needs. Obviously, mastery of the space environment appears crucial in order to cover the various needs in support of the Union's global vocation but still requires a stronger will on the part of its Member States (Legai, 2020, 840–843). A certain consensus is required in Europe among its Member States to face the major space-faring nations acting as a single entity. The concept of the militarisation of space therefore takes on its full meaning when the confrontation that takes place there affects the means dedicated to the defence of certain states. Control of the space environment therefore constitutes, in a way that is perfectly accepted by all, an issue of power, sovereignty and strategic autonomy of the EU or individual Member States. The EU has recognised this with the second-largest public sector space budget in the world after the United States. Europe, at large, strives to promote the establishment and maintenance of space as a global commons for peace and security. A legal approach such as a code of good conduct in outer space promoted by Europe could limit the risks of a militarisation of space, with its possible drifts towards weaponisation, on condition of acceptance and the application by all of such a code.

Notes

1 At the December 2019 Leaders' Meeting in London, Allies declared space a fifth operational domain, alongside air, land, sea and cyberspace. In their declaration, NATO Leaders stated: *"We have declared space an operational domain for NATO, recognising its importance in keeping us safe and tackling security challenges, while upholding international law"* https://www.nato.int/cps/en/natohq/topics_175419.htm.
2 Speech by High Representative/Vice President, Josep Borrell at the 14th European Space Conference, Brussels, 25 January 2022.
3 As of 2019 known deployments of weapons stationed in space include only the Almaz space-station armament (Russian space station program in the early sixties that carried a cannon) and pistols such as the TP-82 Cosmonaut survival pistol (for post-landing, pre-recovery use).
4 The United States, China, Russia, Japan, India and Israel are all investing in hit-to-kill systems to be used for anti-satellite (ASAT) or missile defence. The ASAT capacities of major space powers are already known. China was involved in the most high-profile incident of ASAT testing in 2007, when it used a missile to blow up one of its own defunct weather satellites in LEO. China has also made significant progress in developing anti-ballistic missiles (ABMs) which like ASATs can also be used to target other state's intelligence, surveillance and reconnaissance satellites. Both the US and Russia have also successfully tested ASAT weaponry – the US took down a low-orbit defunct

satellite in 2008 and the Russians completing a flight test of the A-235 Nudol direct ascent ABM. North Korea recently joined the race, launching its own satellite into space. Iran will also be on the way to improving its ASAT capabilities, if it follows through on the 2013 announcement that it is setting up a facility to track orbiting objects.

5 There are currently several types of known space weapons and others are being invented all the time, often secretly. Scientists have already developed directed-energy weapons, such as laser and particle beams to project energy at targets and make them inoperable. Lasers are a cost-effective method for addressing smaller threats, such as shooting down drones or stopping small ships. Lasers also decrease collateral damage, as they are very precise. On-going projects include that of the Defence Advanced Research Projects Agency, which is working on a weapon which will shoot molten metal, with the help of electromagnets – known as Magneto Hydrodynamic Explosive Munition (MAHEM).

6 Manoeuver by the International Space Station to avoid the remains of a Japanese rocket on September 22, 2020.

7 Any progressions made into launching rockets into space are directly relevant to increasing a nation's capability of firing ballistic and intercontinental missiles, or to reach a satellite in Low Earth Orbit (LEO).

8 In accordance with article 7.4 of the *Decision No 541/2014/EU of the European Parliament and of the Council of 16 April 2014 establishing a Framework for Space Surveillance and Tracking Support (SST)*. The top Member States contributing to European SST services are France, Germany, Italy, Spain and UK. Since then, three EU countries joined the SST consortium: Poland, Portugal and Romania. The UK left following the Brexit.

9 Astronaut Rescue and Return of Launched Objects from 1969, Registration of Objects Launched into Outer Space from 1975, State Activities on the Moon and Other Celestial Bodies from 1984.

10 United Nations Committee for the Peaceful Use of Outer Space (COPUOS).

11 The Group of Governmental Experts (GGE) consists of a small group of international space experts from a selection of space faring countries with the main objective to improve international cooperation and reduce the risks of misunderstanding and miscommunication in outer space activities. The GGE meet for the first time in New York, July 23–25, 2012, a second time in Geneva, April 1–5, 2013 and for the last time in New York, July 8–12, 2013.

12 The outcome consensus report was submitted to the 68th Session of the UN General Assembly in 2013 and consists of a set of voluntary TCBMs for outer space activities and recommended for states. In particularly activities on exchange of information between countries space policy and activities, risk reduction notifications and visits by experts to national space facilities. Furthermore, it recommended establishing increased coordination between the Office for Disarmament Affairs, the Office for Outer Space Affairs and other appropriate UN entities.

13 Capital, *Le Président d'Arianespace plaide pour "une grande ambition pour l'Europe dans l'espace"*, 10 September 2018.

Bibliography

Chloé Duffort, 2020, L'espace: y Préserver la Paix, y Prévenir la Guerre, Defense Space Talks 2020, *IEP Bordeaux, La Chaire Défense & Aérospatiale*, 7 Octobre 2020.

Chris Johnson, February 2014, *Draft International Code of Conduct For Outer Space Activities Fact Sheet*, Secure World Foundation.

Crespi Caravella, and Menghini Salvatori, 2021, European Technological Sovereignty: An Emerging Framework for Policy Strategy, *Intereconomics, Review of European Economic Policy* Vol. 56, No. 6, pp. 348–354.

Council, 2016, *Implementation Plan on Security and Defence, Council of the European Union*, 14392/16, Brussels, 14 November 2016.

Council, 2021, *Multiannual Financial Framework 2021–2027 and Next Generation* EU, https://www.consilium.europa.eu/media/47567/mff-2021-2027_rev.pdf

EARSC, 2020, *EU Space Programme*, https://earsc.org/2020/12/21/eu-space-programme-2021-2027.

EDA, 2018, *Exploring Europe's Capability Requirements for 2035 and Beyond, Insights from the 2018 Update of the Long-Term Strand of the Capability Development Plan*, RAND Europe, June 2018, Brussels.

EEAS, 2014, *Draft International Code of Conduct for Outer Space Activities*, 31 March 2014, Brussels.

ESA, 2021, *Mitigating Space Debris Generation*, https://www.esa.int/Safety_Security/Space_Debris/Mitigating_space_debris_generation

ESA, 7 April 2021, *Agenda 2025*, https://www.esa.int/About_Us/ESA_Publications/Agenda_2025.

EU, 2013, *Fourth European Union Draft Code of Conduct for Outer Space Activities*, 16 September 2013.

EU, 2014, *Fifth European Union Draft Code of Conduct for Outer Space Activities*, 31 March 2014.

EU, June 2016, *The Global Strategy for the Foreign and Security Policy of the European Union, Shared Vision, Common Action: A Stronger Europe*, http://europa.eu/globalstrategy/en.

European Commission, 2016a, Communication to the European Parliament, the Council, The European Economic and Social Committee and the Committee of the Regions, *Space Strategy for Europe*, 26 October 2016, COM(2016)705, Brussels.

European Commission, 2016b, *European Defence Action Plan: Towards a European Defence Fund, 30 November 2016*, https://www.europarl.europa.eu/legislative-train/theme-europe-as-a-stronger-global-actor/file-european-defence-action-plan.

European Commission, 2021a, *Defence Industry and Space (DEFIS)*, https://ec.europa.eu/info/departments/defence-industry-and-space_en.

European Commission, 2021b, *Long Term EU Budget*, https://ec.europa.eu/info/strategy/eu-budget/long-term-eu-budget/2021-2027_en.

European Parliament, December 2020, *In-depth Analysis, The European Space Sector as an Enabler of the EU Strategic Autonomy, DG for External Policies of the Union, Policy Department*, Brussels.

Florence Parly, 2019, *Stratégie Spatiale de Défense, Rapport du Groupe de Travail «Espace»*, Ministère des Armées, DICOD, Juillet 2019, Paris.

General Secretariat of the Council, 21 July 2020, *Special Meeting of the European Council (17, 18, 19, 20 and 21 July 2020) – Conclusions*, EUCO 10/20, CO EUR 8, CONCLU 4, Brussels,

German MoD, 13 July 2021, *Germany Establishing New Military Space Command*, https://www.defensenews.com/space/2021/07/13/germany-establishes-new-military-space-command

Jack M. Beard, 2016, Soft Law's Failure on the Horizon: The International Code of Conduct for Outer Space Activities, *University of Pennsylvania Journal of International Law*, Vol. 38, No. 2, 15 February 2016.

Jana Robinson, 2010, *Transparency and Confidence-Building For Space Security and the EU Draft Code of Conduct*, 10–11, EISC Conference, Bucharest, Bloc 3: Security, ESPI. JORF n 0208 du 7 Septembre 2019, *Arrêté du 3 Septembre 2019 Portant Création et Organisation du Commandement de L'espace*.

Louis Perez, 2019, L'application du Droit Des Conflits Armés à l'Espace Extra-Atmosphérique, *Note IRSEM*, 31 janvier 2019.

Michael J. Listner, 2012, Geopolitical Challenges to Implementing the Code of Conduct for Outer Space Activities, *E-International Relations*, 26 June 2012.

Nayef Al-Rodhan, 2018, Weaponization and Outer Space Security, *Global Policy Opinion*, 12 March 2018.

Nicole Gnesotto, October 2009, *What Ambitions for European Defence in 2020? The Need for a More Strategic EU*, EU ISS, second edition, Paris.

Pascal Legai, 2020, *La Notion D'espace Militaire Fait-Elle Sens Dans L'approche Spatiale De l'Union Européenne?* Annuaire Français des Relations Internationales (AFRI), Centre Thucydide, édition 2020.

Pascal Legai, 2021, *Espace: L'état du Droit International Pour y Prévenir ou gérer Les Conflits, La Revue DiplomatiqueLes Nouveaux Défis Juridiques et Géopolitiques du Secteur Spatial*, page 194, Revue trimestrielle n°13, Avril 2021, Institut EGA.

P.N. Tripathi, 2013, Weaponisation and Militarisation of Space, *CLAWS Journal*, Winter 2013.

RAF, 1st April 2021, *UK Space Command, The Security of Space Is Critical to Our Everyday Lives, from Communications, to Transport, to Agriculture*, https://www.raf.mod.uk/what-we-do/uk-space-command.

Reaching Critical Will, 2014, *Outer Space, Militarization, Weaponization and the Prevention of an Arms Race*.

Regulation (EU) 2021/696 of the European Parliament and of the Council of 28 April 2021 establishing the Union Space Programme and the European Union Agency for the Space Programme and repealing Regulations (EU) No 912/2010, (EU) No 1285/2013 and (EU) No 377/2014 and Decision No 541/2014/EU.

Sander Ruben Aarten, 21 April 2020, *The Impact of Hypersonic Missiles on Strategic Stability, Russia, China and the US*, Military Spectator.

UNOOSA, 2019, *Guidelines for the Long-Term Sustainability of Outer Space Activities of the Committee on Peaceful Uses of Outer Space Adopted*, UNIS/OS/518, 22 June 2019.

US Department of Defense, 2004, *National Military Strategy of the United States of America 2004: A Strategy for Today, A Vision for Tomorrow*.

4 From Fragmented Space to the Space University Institute[1]

Thomas Hoerber

The beginnings of European space policy were not easy and certainly conflicted with missiles being used as weapons of terror by the Nazis in the *Blitz* (Reinke, 2004: 21–35). Delayed by these belligerent origins and a lack of funding in the immediate post-war period, European space policy was founded on exclusively peaceful purposes in the 1960s and in the European Space Research Organisation (ESRO) and the European Launcher Development Organisation (ELDO) which, in the 1970s, merged into the European Space Agency (ESA Convention, 1980, Preamble, p.1). These peaceful purposes have been a vitally uniting factor in the supranational integration process of the European Union and its previous organisations, despite the fact that the EU has been a late-comer in European space policy. As outlined in the Convention of the ESA, these peaceful purposes also form the foundations of the intergovernmental European integration of the ESA (Hoerber, 2009a, 2009b). When ESA was founded, the Cold War cast its shadow over the world but allowed Europe to unite to an extent never previously seen in the history of the continent. All major space-faring nations in Europe followed distinct space policies, particularly with regard to accepting US dominance in the field as Germany and Britain did, or by seeking independence, as seen in the French development of Ariane launchers (Hoerber, 2016). However, these different national space agendas materialised within the loose framework of ESA and with the purpose of driving forward European integration (Gibson, 7/5/2014, Paris).

Since WWII, three structures of space technology can be identified which are all classic Cold War space objectives. Some of these are resurfacing in the wake of the COVID19 crisis, which has generally reinforced nationalist and isolationist tendencies.

Firstly, applications, such as satellites and rockets, at which, today the EU excels with Ariane, Copernicus and Galileo, but in which China, for example, also has recently developed major ambitions, such as the use of independent access to space through Chinese-built launchers, a full spectrum of satellite technology, and the beginnings of a Chinese space station (Thomas, 2019). These current applications often originate from military use during the Cold War, such as intercontinental missiles and spy satellites. In Europe, applications were more frequently used for civilian purposes than in the United States, such as for weather forecasting and telecommunications. In most of the rest of the world, however, applications have

DOI: 10.4324/9781003230670-6

never lost their military roots and, even in the EU, the dual-purpose nature of space technology has led to more military applications in European space technology, for example within the framework of NATO and more recently the military arm of the Common Foreign and Security Policy in the European Satellite Centre, formerly, the Western European Union and potentially in the future in the EU Permanent Structured Cooperation (PESCO).

Secondly, space has always held this element of human curiosity about what lies beyond Earth. Stepping on the Moon had the objective of Cold War dominance but also fostered exploration to enhance human understanding. This idealist motive must not be underestimated as a driver of European space policy - for example, fundamental research of the sun - and also seems to be one of the underlying motives for China to go to the dark side of the Moon. Recent objectives of going to Mars have, to some extent, not only revived the Cold War objective of demonstrating excellence, domination and hegemony but also shown a fundamental human drive to go further and to discover what currently lies beyond our comprehension. ESA has become one of the leading institutions in developing our understanding of what we cannot yet comprehend, of fundamental research in this direction and of curiosity which is engrained in human nature.

Thirdly, during the Cold War, space clearly served the purpose of domination and hegemony as captured in the realist theory of international relations (Waltz, 1979; Morgenthau, 1985). More recently, the testing of anti-satellite weapons by a number of nations still shows space as a political tool of domination, threat and posturing. The example most often used in the West is China's anti-satellite weapons test in 2007 (Kan, 2007). However, such sabre-rattling clearly takes us into an area where power politics blur objectivity. Even the European flagship programmes Galileo and Copernicus can be seen as tools of domination; both are clearly capable of fulfilling military functions and are finding more and more interest from the European military-industrial complex (Oikonomou, 2019, 2021), despite the fact that they were conceived exclusively for peaceful purposes (Feyerer, 2016). What can reasonably be said is that both Galileo and Copernicus are not under military command, in contrast to similar services in the United States, for example, the Global Positioning System (GPS). Galileo, in particular, was conceived in a spirit of anti-domination to create European independence after the daunting experience of the Balkan wars in the 1990s during which the United States degraded GPS services, even for its European allies, which posed major problems, particularly for European civil aviation. From this purely realist perspective, space technology is just another tool to increase the power of its owners. With the COVID-19 crisis, such realist arguments have been strengthened. Space does not have to go down that road. Dual-use potential can also mean that space technology is used principally for peaceful purposes, for cooperation and even for understanding among countries that might experience tensions in their relations on Earth. In the past, the International Space Station (ISS) is a good example whereby Russia and the United States worked together, even during times of major friction. Collaboration between China and Europe in space may become another example in the future, if the realist temptation is kept in check in the domain of space.

This spirit of cooperation has been prevalent in Europe from the beginning of its space activities. They were exclusively civilian, of which the two aforementioned European space flagship programmes are good examples, i.e. civilian aspiration and principally civilian use of these satellite constellations. However, space activities, even in Europe, are experiencing an influx of more and more military funding. In most states, this would not be problematic. Particularly during the Cold War, civilian space activities often went hand in hand with military applications. A good example is the development of launchers, which were used not only for space exploration but also as intercontinental missiles that could trigger a nuclear apocalypse. In Europe, this separation of military and civilian use of space technology has been more strongly pronounced, because of the military origins of space technology in WWII, as outlined above. What ESA and the EU have experienced, however, is that in recent years, public funding for space is limited, difficult to maintain, and even more difficult to increase. Therefore, compromises always have to be found. The oldest compromise, which is not limited to Europe, is to interpret "peaceful purposes", as in Art IV of the Outer Space Treaty (1967), as the "non-offensive use of space technology". This is an interpretation which one could widely find outside Europe and which in recent years has also found more friends in ESA and the EU. Reasons for turning "peaceful" into "non-aggressive" can easily be found. In Europe, they range from Russian intransigence with cyberattacks through satellites, to economic competition with China, leading to the more and more popular policy line of hardening up the European space infrastructure for greater resilience against attacks or environmental phenomena (Weiterring, 2019). Finally, everyone who knows a bit about space technology has understood that an object that travels at tens of thousands of kilometres per hour can be peacefully sitting on its orbit or can become a bullet when used in an aggressive way. Galileo satellites can be used not only for geo-localisation signal in a car, but also as positioning signals for military use. Copernicus Earth observation pictures can measure pollution in the atmosphere. They can also be used for military intelligence purposes. This is what is generally accepted as the dual-use nature of space technology.

In Europe, this leads to the question of whether space policy will remain peaceful – a truly European debate which in most other states would be considered purely academic, precisely because of that dual-use nature of space technology. The reason why this debate is so important in Europe comes from the nature of the European integration process, which finds its roots in the experience of WWII and the resulting commitment to peace in Europe (Telò, 2006), both in the EU and ESA. European space assets have dual-use potential, but European space policy has remained civilian, keeping military use at bay, because this tendency of militarisation (Oikonomou, Hoerber, forthcoming; Oikonomou, 2021) seems to contradict Europe's commitment to peace, but it may be seen as another example of the European Union growing up to its responsibilities as a federation and stately actor on the international stage where it may have to defend its assets and interests.

What is important is not to forget the civilian roots of European space policy. There we find the potential for the EU to become something new, instead of most states vying for power in a realist zero-sum game. This potential of breaking new ground should not be underestimated by the EU and ESA. This is even more important, because security arguments have become more prevalent for space use, and this may eventually lead to the weaponisation or even to conflict in space (Leisske, 2017). The founding of space forces is one worrying sign in this direction (Pawlyk, 2018). Currently there are no rules, should this happen. Existing international space law dates from the middle of the Cold War (Outer Space Treaty, 1967). A new treaty seems necessary to better reflect contemporary international politics. In a realist scenario, which we are currently seeing in space, the weaker actors usually have an interest in setting up rules (Leisske, 2017). This may explain why the EU tried some time ago with its Code of Conduct but failed (Mutschler, Venet, 2012). Other medium space actors, such as Japan, India and China, may have similar interests. Building a consensus among them around their interests and sensitivities may become the basis for new space diplomacy.

However, space can provide more. It has always been at the forefront of technological development and innovation which has led to one beacon of hope for humankind, the ISS. It has been a haven of peace. In recent years, ESA has proposed a Moon village in the same spirit (Wörner, 2016). China, which is currently excluded from the ISS, might be able to participate in such an open concept of a Moon village. This collaboration is the source of hope that space can bring to this planet, rather than the militarisation of the space domain which holds the danger of rehashed old power politics which have led to substantial tragedy in the past, not least in Europe. Europe may have a role to play in avoiding such calamities for humankind in the future by developing an inclusive space policy. This may counter the increasingly realist policies resulting from the COVID19 crisis with an idealistic manner of cooperation and effective institutions.

Institutional Setting

In Europe, the debate about bringing together the ESA and EU space policy has been raging for more than a decade (Creola, 2001; von der Dunk, 2003; Hobe, 2004; Verheugen, 2005; Gaubert, Lebeau, 2009; Hoerber, 2009a, 2009b; Peter, Stoffl, 2009). This debate has not really led to a coherent and more effective European space policy, mainly because of the fundamentally different nature of both institutions – in a nutshell, the intergovernmental ESA versus the supranational EU. There are many more facets and details to this description, but it is still the best description of the fundamental difference that exists between these two European space institutions. For that reason, ESA has not become the EU space agency and, since the ESA Ministerial Council Resolution 3 of 2014, reconfirmed in 2016, it has become clear that the Member States of ESA have no intention of bringing ESA into the fold of the EU (ESA, 2014). As a consequence, the EU space agency was founded as a potential competitor to ESA, but more importantly, it was also founded as the agency that exploits and that will potentially implement EU space programmes, most importantly the flagship programmes, particularly Galileo. The result is, however, that the European space institutional landscape has become even more fragmented. Therefore, it seems that there is no satisfactory conclusion to the debate about the institutional setting of a European space policy in an exclusively European context. The EU will want to decide where European taxpayers' money goes in the space sector. With increasing amounts, as has been the case over the past decades, this demand will become more and more important. And ESA will insist on its technical expertise, its longer history and simply its independence from the EU.

What has been misleading in this debate is that in European space policy and the ESA, the common denominator has been seen in the term "European", when what we might actually want to work towards is "Space". Taking "European" out of the institutional equation may actually help to take Europe and the world towards more space policy. The idea that this chapter would like to propose, therefore, is to turn ESA into the Space University Institute (SUI) fostering research, training and innovation, as ESA has always done. Leaving the utilitarian side of downstream usage of space applications to the EU and focusing ESA on what it does best, i.e. fundamental research, training of personnel and space innovation. The EU is better than ESA to reap the benefits of Galileo and Copernicus and to provide them as raw data suppliers or applications to European entrepreneurs. The EU has the legitimate mandate to provide and demand maximum benefit from these technologies for European citizens. ESA, however, is better than the EU in what it has always done, fundamental research, training and innovation in space. If this analysis is correct, the most constructive way forward may be to further open up ESA internationally, as is already the case with Canada which is an associated Member State. This holds the potential of inviting all space-faring nations to a common space policy. In such a setting, ESA would inevitably retain its intergovernmental nature, because other states would only join on that basis. Supranationalism could remain an EU phenomenon. A blueprint for such intergovernmental collaboration may be the positive experience Europe has had with the European University Institute (EUI), which has fostered excellence in research, training some of the brightest minds in Europe, and in innovation in the widest sense. All that has happened on the basis of an intergovernmental agreement founding the EUI. The EUI has nevertheless had a certain federating effect for participating Member States, clearly adding to European integration, just like ESA has done. Taking ESA international on the same basis may be an interesting way forward for bringing together capable and willing partners around space projects which may add up to a more ambitious space policy for humankind. Europe might be seen as the honest broker who can host such world space policy. A recent proposal by the Green party EFA (2021) in the European Parliament suggested a "European Space Academy". Take this idea of a training and research institution to an international level and one may find that space can be a uniting bond for humankind, as the ISS has been, for example. Evidently, this would avoid the danger of renationalisation and further isolation, for example of China, as a consequence of the COVID19 crisis. At the very least, it will be a gesture towards those who want to progress towards a common endeavour into space. Such a SUI would allow the federation of world competences in space technologies. It would also allow the EU to develop a truly European space policy serving its citizens. And it would allow ESA to progress and fulfil its purpose of fostering space policy. Putting the nominal headquarters of the EUI on the Moon, with an aspiration to go there and stay there, would give it a real objective. The SUI would be situated in a neutral setting and show ambition which may hopefully draw in other space enthusiasts, may they be nations, companies or individuals. The Moon village concept of Jan Wörner may serve as a preliminary vision for what we cannot yet imagine (Wörner, 2016; Wörner, Foing, 2016; Köpping Athanopoulos, 2019).

Note

1 This chapter was first published as Hoerber, T. 'From fragmented space to the Space University Institute', in: *Journal of Chinese Economic and Business Studies*, DOI: 10.1080/14765284.2022.2081486, May, 2022.

Bibliography

Primary Sources

EFA (2021) Green European Space Policy, forthcoming as proposal in the European Parliament.
ESA Convention (1980) *ESA Communications*, Noordwijk.
ESA (2014), Resolution 3.
Outer Space Treaty (1967) available at: https://outerspacetreaty.org/, accessed on 15 July, 2021.

Secondary Sources

Creola, P. (2001) 'Some comments on the ESA/EU space strategy', *Space Policy*, Vol. 17, pp. 87–90.
Feyerer, J. (2016) 'Lessons from Galileo for future European public-private partnerships in the space sector', in: Hoerber, T. & Stephenson, P. (eds.) *European Space Policy - European Integration and the Final Frontier*, Routledge, London, Chapter 13.
Gaubert, A., & Lebeau, A. (2009) 'Reforming European space governance', *Space Policy*, Vol. 25, pp. 37–44.
Gibson, R. (7 May, 2014) *Speech at the 50th Anniversary Celebration of ESA*, Paris.
Hobe, S. (2004) 'Prospects for a European space administration', *Space Policy*, Vol. 20, pp. 25–29.
Hoerber, T. (2009a) 'The European Space Agency (ESA) and the European Union (EU) – The next step on the road to the stars', *Journal of Contemporary European Research (JCER)*, Vol. 5, No. 3 pp. 405–414.
Hoerber, T. (2009b) 'ESA + EU, ideology or pragmatic task sharing?' *Space Policy*, Vol. 25, pp. 206–208.
Hoerber, T. (2016) 'Consolidation or Chaos – The way to ESA', in: Hoerber, T. & Stephenson, P. (eds.) *European Space Policy - European Integration and the Final Frontier*, Routledge, London, Chapter 1.
Hoerber, T., & Forganni, A. (eds.) (2021) *European Integration and Space Policy – A Growing Security Discourse*, Routledge, London.
Hoerber, T., & Lieberman, S. (eds.) (2019) *A European Space Policy – Past Consolidation, Present Challenges and Future Perspectives*, Routledge, London.
Hoerber, T., & Sigalas, E. (eds.) (2017) *Theorising European Space Policy*, Lexington Books, New York.

Hoerber, T., & Stephenson, P. (2016) *European Space Policy - European Integration and the Final Frontier*, Routledge, London.

Kan, S. (2007) *China's Anti-Satellite Weapon Test, CRS Report for Congress*, available at: https://fas.org/sgp/crs/row/RS22652.pdf. Accessed on 15 July, 2021.

Köpping, Athanopoulos. (2019) 'The moon village and space 4.0: the "open concept" as a new way of doing space', in: Hoerber, T. & Lieberman, S. (eds.) *A European Space Policy – Past Consolidation, Present Challenges and Future Perspectives,* Routledge, London, Chapter 9.

Leisske, M. (2017) 'Power politics and the formation of international law: A historical comparison', in: Hoerber, T. & Sigalas, E. (eds.) *Theorising European Space Policy,* Lexington books, New York, Chapter 6.

Morgenthau, H. J. (1985) *Politics Among Nations: The Struggle for Power and Peace,* Knopf, New York.

Mutschler, M., & Venet, C. (2012) 'The European Union as an emerging actor in space security', *Space Policy*, Vol. 28, No. 2, May 2012, pp. 118–124.

Oikonomou, I. (2019) 'The socio-economic logic of the EU space strategy', in: Hoerber, T. & Lieberman, S. (eds.) *A European Space Policy – Past Consolidation, Present Challenges and Future Perspectives,* Routledge, London, Chapter 2.

Oikonomou, I. (2021) 'The strategic utilisation of the US in EU space policy discourse', in: Hoerber, T. & Forganni, A. (eds.) *European Integration and Space Policy – A Growing Security Discourse,* Routledge, London, Chapter 2.

Oikonomou, I. Hoerber. (eds.) (forthcoming) *The Militarization of Space Policy: The EU and Beyond*, Routledge, London.

Pawlyk, O. (2018) 'It's official: Trump announces space force as 6th military branch,' *Military. com*, 18 June, 2018, available at: https://www.military.com/daily-news/2018/06/18/its-official-trump-announces-space-force-6th-military-branch.html. Accessed on 15 July, 2021.

Peter, N., & Stoffl, K. (2009) 'Global space exploration 2025: Europe's perspectives for partnerships', *Space Policy*, Vol. 25, pp. 29–36.

Reinke, N. (2004) *Geschichte der Deutschen Raumfahrt*, Oldenburg Verlag, Munich.

Telò, M. (2006) *Europe: A Civilian Power? European Union, Global Governance, World Order*, Palgrave Macmillan, Basingstoke.

Thomas, A. (2019) 'China's cooperation with Europe – The supporting public narrative of space exploration in China', in: Hoerber T. & Lieberman, S. (eds.) *A European Space Policy – Past Consolidation, Present Challenges and Future Perspectives*, Routledge, London, Chapter 1.

Verheugen, G. (2005) 'Europe's space plans and opportunities for cooperation', *Space Policy*, Vol. 21, pp. 93–95.

von der Dunk, F. (2003) 'Towards one captain on the European spaceship – Why the EU should join ESA', *Space Policy*, Vol. 19, pp. 83–86.

Waltz, K. (1979) *Theory of International Politics*, McGraw-Hill, New York.

Weiterring, H. (2019) *France Is Launching a 'Space Force' with Weaponized Satellites*, 2 August, 2019, available at: https://www.space.com/france-military-space-force.html. Accessed 15 July, 2021.

Wörner, J. (2016) 'Moon village: A vision for global cooperation and space 4.0', *European Space Agency*, available at: http://blog.esa.int/janwoerner/2016/11/23/moon-village/. Accessed 23 January, 2018.

Wörner, J., & Foing, B. (2016) 'The "moon village" concept and initiative', *Annual Meeting of the Lunar Exploration Analysis Group*, available at:http://www.hou.usra.edu/meetings/leag2016/pdf/5084.pdf. Accessed 18 March, 2019.

Part II
Actors, Issues & Interests

5 Outer Space, Debris and the Militarisation of Space

Isabelle Sourbès-Verger

The issue of space debris has to be considered at the global level, as it is a growing challenge for the international community. Only in the last 20 years, Europe has become aware of the necessity to deal with the subject, primarily with the goal of protecting its satellites. The accumulation of launches since the beginning of the space era – as many as 7,500 objects by January 2017 (UN Office for Outer Space Affairs, undated) – has resulted in the multiplication of uncontrolled objects in orbit: launcher stages, defunct satellites, small fragments due to the opening of the launcher's nose cone, and various materials due to processes of deterioration.[1] Even if technical measures were adopted by space agencies to limit its increase, the number of debris keeps rising, albeit at a moderate speed. The concerns related to space debris and the risks of collision with operational spacecraft are thus increasingly pressing in light of the intensification of space activities, including the expected multiplication of small satellites.

Another security challenge, though of a different nature, is that European governments and their armed forces have to consider the potential weaponisation of outer space (Pellegrino & Stang, 2016: 23–29). The risk of war in outer space would further fuel the proliferation of space debris. In this context, the possibility of anti-satellite (ASAT) weapons testing and, more specifically, of ground-based missile launches targeting satellites is particularly worrying, as these generate thousands of debris (Lele et al., 2012: 35). After years of an American and Russian moratorium on ASAT physical tests, the 2007 ASAT test made by China coupled with the systematic US rejection of a United Nations ban on space-based weapons (Joseph, 2006) has made the discussion around the security of space assets an increasingly sensitive and debated subject at the international level.

Europe takes part in these discussions on several accounts, depending on the nature of the actor involved, i.e. the Member States (MS), the European Space Agency (ESA) or the European Union (EU) Council and Commission. With regard to the debris issue, national space agencies and ESA have played a significant role since 1993 in the Inter-Agency Space Debris Coordination Committee (IADC), an international forum for coordination on policies, guidelines and standards relating to space debris mitigation. More recently involved in this field, the EU proposed a diplomatic initiative in 2008: the International Code of Conduct for Outer Space Activities (ICoC). It aims at enhancing the safety,

DOI: 10.4324/9781003230670-8

security and sustainability of space activities by promoting transparency and confidence-building measures. It is important to take into account that contrary to their American counterparts, European national armed forces have a more limited number of military satellites dedicated to operational purposes. Indeed, these questions have become of military concern quite recently and only for some MS.

This chapter first examines how the improvement in European space capabilities has led a growing number of actors within Europe to develop a stake in the debris issues. It then describes the pro-active approach developed by ESA and the EU in the field of space surveillance since the 2000s. It concludes with an analysis of the current status of the Space Situational Awareness programme, setting it in the European and international context. The topic of space debris demonstrates the complexity of Europe's situation when dealing with new security challenges at the levels of MS, ESA and the EU. It highlights not only the limits of the European tools compared to these of the other members of the Space club but also the growing awareness of the necessity to develop a common policy and to set up the means for an independent expertise on an increasingly sensitive issue.

Europe's Growing Engagement in Satellite Tracking Issues

In the context of the Cold War, the building of space capabilities of European states benefited from a strong cooperation with the United States. European states relied specifically on the American technical means for specific and complex tasks such as the tracking of satellites. Due to this support, it took time for Europe to realise that the issue of space surveillance could be a crucial element for its position on the international scene. This section will detail chronologically the development of Europe's capacities and their limitations compared to those of the US and of the Soviet Union.

Tracking: A Strategic Competence for the Two Superpowers During the Cold War

At the beginning of the Space Age, being able to monitor their own satellites was a significant and unresolved technical challenge for the United States and the Soviet Union. The original US system, established in the context of the International Geophysical Year in 1957–1958, consisted of a Navy radar network made up of low-cost antennas and based on the interferometric measurement of the radio emissions of satellites. The unnoticed launch of Sputnik in 1957 was a shock. It showed that the first American facilities were inadequate to detect and identify space objects whose orbital parameters were unknown. As Soviet satellites were seen as potential threats by the United States, the priority was to follow them and acquire a trajectography allowing the interception of enemy satellites. This was done by combining the data acquired by the Navy, the US Air Force (USAF) and the Army (Shepherd project) (Arpa, 1959: 27). By 1961, the North American Aerospace Defense Command (NORAD) Combat Operations Centre was made responsible for controlling a large Space Detection and Tracking System (SPADATS). While it would later become the US Space Surveillance Network (SSN), it showed unparalleled performances from the start (Stares, 1985: 131–133).

The Soviet Union followed a similar approach, though slightly later. The monitoring of the first satellites also relied on fairly basic cameras. The foundations of space-based monitoring were laid in 1962 as part of the space surveillance and ASAT programmes. The global tracking project for objects in orbit began in 1963 and was approved in 1965, giving birth to the Tsentr Kontrolya Kosmicheskogo Prostranstva (TsKKP),[2] which consisted of radar and optical means controlled by the Tsentr Upravlyeniya Polyotom (TsUP), located near Moscow. The system was considered operational in the mid-1970s, even though it was less effective than its American system with respect to the precise identification of the type of satellite being tracked and the detection of geostationary satellites.

Accordingly, during the Cold War, the initial concerns of the two superpowers in the field of space activity monitoring were part of security issues typical of the era and especially of nuclear deterrence approaches. The beep of Sputnik was considered by the United States as a true threat to their national security, not only because of the ideological impact of a "Soviet first" but also because of the anxiety around a potential bomb on board of the satellite that was seen as a sword of Damocles able to reach any place of the territory without notice. On the Soviet side, the tracking of the US satellite orbitography was also of the utmost interest in order to avoid US reconnaissance satellites to acquire information on strategic and potentially vulnerable ground targets. To a larger extent, both superpowers sought to acquire the capability of tracking down a greater deal of objects because of the increasing number of satellite launches. This effort aimed also at developing a first step of space traffic management practices. Given that Europe was under American nuclear protection, this issue was not of urgent concern.

Europe in Space: A Civilian Field of Activity

In the 1960s, mastering launcher technology in parallel to nuclear capability was perceived as a symbol of sovereignty that helped also improving international standing. On this matter, France and the United Kingdom (UK) were the first European states to nurture space ambitions and develop their own satellites in parallel to a programme of small launchers[3] that stemmed directly from their missile programmes. However, their space programmes were limited by the scarcity of their financial and technical means. The importance of the technical support provided by the United States to implement them is undeniable. In fact, all the national capabilities of European states had been developed thanks to cooperation with the NASA (National Aeronautics and Space Administration). None of them had neither the ambition nor the motive to acquire their own space surveillance means excepted for their own satellites, contrary to the United States and the USSR, for whom space was de facto part of the national interest. This can be illustrated by the UK government's disinterest in the outstanding results of a radio-amateur team of the Kettering College that had been tracking Soviet satellites since the mid-1960s (see Davis, 2016). Yet, while this was ignored at home, the head of the team, Geoffrey

Perry, soon gained recognition in the United States, where he was employed as an expert in charge of the orbit determination of Soviet satellites.[4] In the same vein, most of the expertise developed in Europe during the 1960s and 1970s in terms of tracking of foreign satellites was due to the personal involvement of amateurs participating in an international network of volunteers receiving signals and observing satellites.

In parallel with the first and limited programmes of France, the UK and, to a lesser extent, Germany and Italy, it was early recognised that the scientific, technological and financial challenges of space access and the development of operational satellites could only be addressed through intra-European cooperation (Krige & Russo, 2000: 263–271). Indeed, the scientists committed themselves to creating the European Science Research Organisation (ESRO) in 1962. In this context, the first European Space Data Centre was created as early as 1963, using the American data to conduct the orbit calculations of first European satellites.

Three main arguments were at the core of the European approach to space issues: the political significance and strategic advantage of competences in an advanced technological field, the direct and indirect economic benefits related to the mastery of space activities, and the importance of Earth monitoring programmes. But the issue of space surveillance was not even mentioned, except for the monitoring of the orbits of the national and European space assets. The ESA, a new research and development agency, was set up in the mid-1970s for peaceful purposes. Cooperation was based on an intergovernmental model and primarily driven by scientific interest.[5] The founding convention of the ESA, signed in 1975 by all the members of the organisation, rules out military programmes from the Agency's activities. To that extent, the security issues for the ESA are limited to a technical approach to the threats associated with debris. The European Space Operations Centre (ESOC) took the succession of the former European Space Data Centre with gradually improved technical capabilities. Located in Darmstadt, the Centre has been in charge of controlling ESA spacecraft in orbit and managing the ESTRACK stations, which comprise nine stations in seven countries. This network's essential task is to communicate with satellites in orbit and ESOC is able to support at least 20 missions operates in close collaboration with the US SSN.

In essence, ESA's involvement in research, technology and operational aspects related to space debris has been developing since the mid-1980s. Its first report on debris dates back to 1988 (European Space Agency, 1988) and resulted in a Council Resolution in 1989 on the risks faced by satellites, particularly manned flights conducted in the framework of European participation in the International Space Station program. The recommendations it gave called for an effort to minimise the creation of space debris, as well as acquire the data necessary to assess the problem and its consequences through civilian national facilities. Following the same logic of this wake-up call, ESA became one of the founding members of the IADC in 1993. In 1998, it was joined by the British National Space Centre (BNSC), the French National Space Center (CNES), the German Aerospace Centre (DLR) and the Italian Space Agency (ASI). In this context, it developed competences that would later contribute to the elaboration of the 1998 Space Situational Awareness proposal (discussed below).

A Very Limited Interest from a Few States

The political dimension of security in space is mainly related to the sovereignty of each state and their respective national space policies. A growing interest in the latter has been gaining momentum in the 2000s, although at different levels and according to the MS' own priorities. For instance, the historical proximity between nuclear and space cultures and especially the concept of strategic independence played a significant role in France's decision to develop its own military space capabilities, particularly in the field of Earth observation. Helios was initially intended to provide independent information at a political strategic level (Verger et al., 2003: 341–343). Yet this ambition remains rather limited and does not include space-monitoring capabilities in spite of the lack of autonomy that this implies. Hence, when the first satellite of optical reconnaissance of the Helios family was launched in 1995, the CNES team unsuccessfully tried to establish contact with the small Cerise satellite launched at the same time. They had to wait for the US to inform France of the loss of their satellite due to a collision with a rocket stage. The financing of new generation Helios-2 and the development of a dedicated telecom satellite made space a significant element of French national military capabilities leading to a gradual awareness of the challenges of satellite security in space.

The existence of growing national competences contributes to a more acute perception of potential risks run by satellites in orbit and to the development of some national facilities in the 2000s. This is emphasised by the diversification of the missions operated by these satellites. These include not only Earth observation but also telecommunications and other specifically military missions, such as early warning and Signal Intelligence (SIGINT). In that context, it becomes urgent for European states to consider the risks and threats to the space environment and thus to their military space means. This awareness has been accelerated by the US National Space Policy of 2006, released on August 31, 2006 by President George W. Bush, which reinforces the need to protect space assets against what they see as growing threats from foreign countries, including rogue nations.

The Withdrawal of the European Community

It took some time until the European Community began to play a role in the field of space activity. As far as the debris and safety issues are concerned, the debate on space security (Pasco, 2009: 6) did not take place before the 2000s. It would take an additional ten years before the first commitment took place in matters of space security, with the draft proposal of the Code of Conduct for Outer Space. This delay is due to the fact that at the community level, the priority given to economic development and integration at the beginning of the European construction process proved utterly unsuited to the integration of space ambitions and the beginning of the European Commission's interest in space programmes was mainly driven by a customer approach.

The interest in space security matters was later constructed within another context. The European Single Act established in 1988 conferred a broader mission upon the European Community with regard to political and economic aspects of security. In that context, the need for a collective legislation on space debris was mentioned as a matter of general interest in the 1988 paper "The Community and Space, a coherent approach", drawn up at the time at the request of the European Parliament (European Commission, 1988: 24). In addition, the existence of the second mainstay of European policy, the Common Foreign and Security Policy (CFSP), which came into being with the Maastricht Treaty in 1993, meant that the whole range of possibilities offered by space could be taken into consideration. At last, the European Security and Defence Policy became part of the CFSP with the signing of the Treaty of Amsterdam in 1999. In this framework, space assets could be taken into consideration to support security issues such as crisis management, conflict prevention and humanitarian intervention, known as the Petersberg tasks, which are at the heart of the security challenges as identified by Europeans after the Cold War.

Similarly, in 1999, European ministers called upon the European Commission and the ESA executive to set up a "Coherent European Strategy for Space", an initiative also supported by the European Parliament (European Commission, 2000). In that framework, the decision to develop dual systems, such as the Galileo navigation programme and the Earth monitoring programme GMES (Global Monitoring on Environment and Security), marks a turning point for the EU awareness of the hazards due to natural and man-made debris. This is the outcome of a joint effort deriving from the synergy among the historical actor of European Space Policy, ESA and the European Commission.

New Approach to Security Challenges in Europe in the Mid-2000s

Different reasons can be put forward to explain the new approach of Europe concerning space security issues. Ambitious new programmes were put in place and the attitude of the ESA evolved to give more attention than before to dual-use issues. In parallel, space issues found a new place in the political agenda of the EU taking into account security concerns in the framework of the new treaties (Treaty of Nice, 2001; Treaty of Lisbon, 2007).

ESA and Space Surveillance

It is through the question of debris, which is at the core of its competence, that the ESA Council adopted in December 2000 a resolution for a European Policy on the protection of the space environment. The work was coordinated by the ESOC and focused on the definition of a standard for the safety of orbiting satellites. A workgroup made up of members of the ESA and the national space agencies of Italy, the UK, France and Germany elaborated and presented a project for a European standard in 2004. This standard comprised preventive measures and introduced the principle of orbit protection. It pertained to the conception and production of satellites and launchers, to the operations phase and to the solving of problems posed by these ageing satellites (United Nations Office for Outer Space Affairs, 2004: 2–10).

In parallel to these concerns, and in light of the increasing interest of MS, ESA is eager to develop its own competences and stated new ambitions for space surveillance according to its R&D mission (while the EU is responsible for diplomatic and political initiatives). According to the study on the "Feasibility of performing space surveillance tasks", a space-based optical architecture was proposed in 2005 (Flohrer, 2011: 1029–1042). The main work at that time, however, was that of the European Coordination Group on Space Debris. Its members coming from the BNSC, Centre National d'Etudes Spatiales, Deutsches Zentrum für Luft- und Raumfahrt and ESA carried out a report entitled "Europe's eyes on the sky" (Klinkrad, 2008: 42–48) using studies already issued by the Space Surveillance Task Force such as "Space surveillance for Europe – a Technical Assessment" released in 2006 (del Monte, 2015: 1–6). In the preface, it defined space surveillance as the detection, correlation, characterisation and orbit determination of objects in space. The introduction made it clear that Europe had no systematic operational capability for space surveillance and was strongly dependent on external information, mainly for the US SSN. Calling for the development of an independent system, the report was intended to provide material for an interagency and intergovernmental discussion in a future European Space Surveillance System (ESSS) and, eventually, for a Space Situation Awareness System (SSA).

The results, endorsed by the ESA Cabinet meeting of November 2008, led to the launch of the SSA, implemented as an optional programme with 14 MS participating financially. It focused on three main areas: space weather (SWE), near-Earth objects (NEO) and Space Surveillance and Tracking (SST). Its aim was to give Europe an independent capability to watch for objects and natural phenomena that could harm satellites in orbit. During the 2009–2012 Preparatory Phase of the SSA programme, precursory applications were developed to serve as a test bed for the novel techniques and algorithms needed for the Space Surveillance Tracking System. Although ESA is aware of what was politically at stake with the European SSA due to the sensitive nature of data exchange, the MS with the most capabilities stayed in the background. At the 2012 ESA ministerial Council in Naples, France did not confirm its commitment to the project, while the UK and Germany chose to get involved in the SWE and NEO segments, both much less sensitive. The second phase (2013–2016) has been extended to 2019.

Facing a Broadening Range of Security Challenges: The EU's New Missions on Space Matters

The increasing involvement of the EU in security matters begins in the 2000s. Its role in global security issues increases in parallel to the deep changes implemented by the European institutions. As far as space is concerned, the following chronology highlights the main steps of this new approach. In 2001, the Torrejón

Satellite Centre – created in 1991 under the auspices of the Western EU – became the SatCen (European Union Satellite Centre), an Agency of the Council of the EU (Council Decision, 2014). It was considered an essential asset for the strengthening of the CFSP, especially in crisis monitoring and conflict prevention, as it provided products and services resulting from the exploitation of data including satellite imagery. For the first time, the EU was thus directly affected by the potential vulnerability of a source of information of great importance for the conduct of its foreign policy, under the political supervision of the Political and Security Committee.

In 2003, the European Council laid down the principles of a European security strategy (Council of the European Union, 2003), while the European Commission launched a "Preparatory Action for Security Research" in 2004, with the ambition to later implement a specific programme for research and development in the field of security (European Commission, 2004). In the context of that time, the Report of Experts on Space and Security, also known as the SPASEC report (Space and Security Panel of Experts, 2005), was published in March 2005. It referred both to the "ESDP (European Security and Defense Policy) and Space", a document approved by the Council in November 2004 and to the future "European Space Programme" being elaborated. First, the experts introduced their work with an analysis of European space-based security needs, stressing the need for an independent information capability and giving concrete recommendations regarding the future European space programme so as to effectively contribute to the CFSP's objectives and the ESDP's needs. They also recommend moving closer to the European Defence Agency and, above all, they promote the establishment of global action, the "Coherent European framework initiative". The need for space surveillance capabilities came in second position among the identified capability gaps, right after the necessity for the development of a dedicated architecture taking into account the various national space observation systems. It was clearly specified that non-security-related communities and security/defence communities were involved.

During this period, France develops its own space surveillance system in the framework of an experimental program conducted by the ONERA and placed under the Air Force's responsibility (see below). At the European level, the military community clearly showed its interest in dealing more specifically with the identification and assessment of attacks in space as early as 2006. Yet, two reports, "Outline of Generic Space Systems Needs for Civilian Crisis Management Operations" issued by the Committee for Civilian Crisis Management (2006) and "Weapons in Space", by the WEU's Technological and Aerospace Committee (2007) were not considered by the EU to justify its role in this new field of activity.

China's ASAT test in January 2007 followed by the US one in February 2008 warned the world that voluntary destruction of assets might become a real possibility. The vulnerability of space assets then extended beyond the debris issue. In this respect, the Parliament explicitly criticised the fact that the weaponisation of space was not taken into account among the topics of interest in the 2007 European Space Policy. The Resolution on Space and Security adopted by the European Parliament in 2008 stressed the fact that space assets were necessary for Europe to fulfil its security missions in terms of assessment, ensuring independence in its decision-making process. The Parliament notably advocated the promotion of binding international agreements in order to guarantee security in space (European Parliament, 2008). Therefore, the Commission supported the acquisition of space surveillance skills through the funding of projects more specifically related to services as part of its R&D programme (FP7, 2007–2013).

In fact, the risks about security in space were finally taken into account in a new institutional context – the signing of the Treaty of Lisbon in 2007. For the EU, it meant a new political legitimacy and an opportunity to deal with security matters with largely economic aims, while the purely military aspects remained the responsibility of MS. However, the arguments put forward were mainly of an economic and industrial nature, emphasising the need to protect investments in space infrastructures, to ensure the continuity of services and the viability of space activities and, more broadly, to support the competitiveness of the European industry. Thus, in 2010, the EC asserted that implementing a specific programme was necessary (see EC/ESA, 2010). This policy continued through the framework programme Horizon 2020. It highlighted the objectives of gaining technological independence as well as acquiring the necessary data for space monitoring. So were security arguments in the broadest sense, including the risks and responsibilities of launches, the uncontrolled re-entry of objects and the presence of debris in the most used orbits.

The Code of Conduct: An EU Initiative

The elaboration of a diplomatic proposal, the ICoC, reflects the new European desire to politically contribute to cooperative security in space. While the United States and Soviet Union had reached a kind of consensus on a de facto moratorium on ASAT systems linked to the 1972 Anti-Ballistic Missile Treaty, Russia's continued decline during the Yeltsin era and the US crucial reliance on space assets as a part of their national interest led the USA to give itself the mission to ensure space control by developing weapons in space if deemed necessary. This attitude leads to a new way of thinking of space as a future battleground, with a potential arms race lying ahead as demonstrated by the China ASAT test of 2007. In this context, the code of conduct promoted by Europe aims to offer both a new approach and an alternative to Russia's and China's proposal. In fact, their latter joint proposal on Treaty on Prevention of the Placement of Weapons in Outer Space and of the Threat or Use of Force Against Outer Space Objects (PPWT) submitted to COPUOS since February 2008 remains firmly opposed by the United States.

From the European perspective, the ICoC initiative is a turning point as it expresses a concern that is common to all the players, some of which have only recently begun to take interest in the topic of security in space. These "rules of the road in space" promoted a greater transparency and a more concrete regulation through the COPUOS Technical and Scientific subcommittee. Its ambition was to set in motion a debate that had been stuck on the question of armament by considering a broader perspective on security, one that centred on less controversial topics such as debris (Brachet, 2016: 2–9). The proposal fell within the tradition established by the 1967 Outer Space Treaty and displayed an ambition to preserve the peaceful and sustainable use of space for present and future generations, in a spirit of greater international cooperation, collaboration, openness and transparency.

The contents of the Code were unrestrictive as it required respecting existing treaties and principles, and implementing measures aimed at minimising the risk of collision and interference with other objects or space activities as well as the creation of new debris. It also called for refraining from actions that could endanger or destroy objects in space and accepting transparency and confidence measures (TCBM) such as launch notifications and base visits. This proposal raised criticism on several accounts: it did not clearly forbid space attacks, although the right to self-defence was explicitly mentioned, it did not express a common approach on the aim of space activities, and it had no biding dimension (Lele, 2012). A second version published in March 2014 took this feedback into account. It clarified the aim of the initiative: to guarantee the "security, safety, and sustainability of space activities", making an explicit reference to the prevention of an arms race. However, the procedure conducted outside the UN framework remained a problem. The new consultation that opened in 2016 finally resulted in an admission of failure. Although a European consensus was reached and interest was expressed by some players, the project failed to convince. Nevertheless, the process of having European autonomous monitoring assets had been set in motion.

Towards a New Synergy between MS and the EU?

In Europe, a turning point was reached in 2013 with the Framework for SST Support proposed by the Commission and adopted by the EP and the Council on 16 April 2014 (Council of the European Union, 2014). The EU – through the Commission – would manage the initiative and support it with funds from the R&D budget Copernicus (formerly known as GMES), Galileo and internal security budgets. The aim is to be able to provide European services based on the use of existing data and national sensors without considering developing new sensors. The end users are the MS, the Council, the Commission, the European External Action Service, public and private owners and operators of space vehicles, and the authorities responsible for civil protection. Military users are not explicitly mentioned in the document but are not ruled out either. They may benefit from civil/dual-use services, such as anti-collision, alerts and detection. A consortium was responsible for the establishment, operation and improvement of sensors; the processing of the data; the financing of new sensors if needed; and the implementation of the data policy (Caleb, 2015). On 17 June 2015, by signing the SST consortium agreement, France, the UK, Germany, Italy and Spain laid the foundations for a concrete cooperation system in this field named "EU SST" (see European Commission, 2021).

In collaboration with the consortium, SatCen, which already participated in research projects funded by FP7 and H2020, acted as an interface and service provider for the Commission. It was put in charge of building the support framework and implementing it; identifying and managing risks; ensuring the update of the end users' needs; defining the main guidelines for the governance of the support framework; facilitating the greatest possible participation by MS; and passing on any relevant information to the European Parliament and Council.

The European Space Surveillance Capabilities

Mastering Space Surveillance and its Technical Requirements

Space surveillance is a matter not only of civil but also of military capabilities. The technology developed to survey activities in space thus serves both purposes. Yet it is an important challenge, as it requires a large range of technical means on the ground and potentially in space as well, like software and space data catalogues. The question of debris embeds itself in this global context. A comprehensive space surveillance system must achieve four main missions: monitoring aimed at detecting space objects without any prior knowledge, the determination of orbit parameters and their regular updating as precisely as needed, the identification or characterisation of the object, and targeting if it is intended for an ASAT capability. There are multiple natural constraints: several objects may move across the local sky simultaneously, time intervals between two appearances may vary greatly depending on the satellites, various disturbances can affect the orbits so that the description and the accurate anticipation of the trajectography are complex and require considerable analysis and data processing means. Actually, the system is all the more effective when it can use information acquired by a network consisting of locations that are geographically scattered across the globe.

Various systems are used on the ground and in space: radars have wide fields of view, they are not dependent on weather or lighting conditions, but they are also generally limited to an altitude that does not allow for the tracking of objects in geostationary orbit. They are usually complemented by ground telescopes, which are more effective with perigee altitudes beyond 5,000 km as they detect the sunlight reflected by objects. Finally, space assets (radars, telescopes and lasers) can provide a third source of information that can improve the quality of the data. This quality obviously depends on data processing, whose role is critical in establishing and updating catalogues of objects. Lastly, various levels of space surveillance capabilities may be identified. There are dedicated assets who specifically monitor, count and identify objects in orbit; collateral assets contributing to surveillance even though it is not their main mission; and contributor assets, which may be used occasionally although it is not their usual purpose.

Only the United States and Russia have an operational system today, with a routinely updated space objects catalogue, although at different levels. The American structure is unique in terms of capabilities and in the way the notion of global space surveillance architecture was formalised. The first aim was not only to preserve military satellites' increasingly crucial contribution to the technological superiority of the U.S. forces, but also to take into account the concerns of commercial systems (particularly those dedicated to telecommunications and observation), which are essential tools for the information society. As from August 30 2016, the Joint Space Operations Centre (JSpOC) has been tracking more than 17,000 objects orbiting Earth: 4,232 are payloads or satellites (850 active) and 13,583 are rocket bodies and debris.[6] Using a worldwide network of 30 space surveillance sensors (radars and optical telescopes, both military and civilian), the SSN allows to observe the current orbiting objects and to catalogue and update the position and velocity of each. These update from the Satellite Catalogue, a comprehensive list of the numbers, types and orbits of all trackable objects in space.

Although its performances are not as thorough, the Russian space monitoring system (SKKP)[7] under the responsibility of the Aerospace Defence Troops (EKR in Russian) has been considered a crucial part of Russia's security architecture since 2000. It is now able to compile and update a catalogue of over 5,000 space objects. Over the coming years, more than ten new generation lasers, optical and radar systems are expected to dramatically improve the Russian network.

European Capabilities

Compared to the US and Russia, Europe's own resources for space surveillance are quite low. They are shared between several players, the main ones being two MS – France and Germany – and the ESA network. The process of data exchange is a key issue that remains unsolved due to the lack of global political leadership.

The GRAVES radar (Grand Réseau Adapté à la Veille Spatiale), specifically designed for space surveillance, is a special division of the Commandement de la Défense Aérienne et des Opérations Aériennes (CDAOA). It is a bistatic radar, i.e., with a transmitter and a receiver in different locations. It performs electronic scanning and continuous transmission in VHF band by using the 400-km distance between the transmission location in Broyes-lès-Pesnes, near Dijon, and the receiver system in Apt, on the Albion plateau. The system uses Doppler detection and advanced signal processing as a special effort was made to develop software converting raw measures coming from a single sensor (rather than a geographically distributed network) into a database of orbital parameters. The radar was considered operational in 2005 and now has an updated database of over 2,000 objects.[8] A development of GRAVES 2, envisioning its use until 2025–2030, is planned in accordance with the plans of the military program law of 2014–2019.

The German radar TIRA (Tracking and Imaging Radar) has very different characteristics. It is a civilian facility operated by the Research Establishment for Applied Science (FHR in German), in Wachtberg, near Bonn. It is able to track objects in the L Band and has a Ku-band imaging capability that complements the data provided by the US catalogue. It is also used to help identify the objects spotted by GRAVES.

Additional capabilities can also be obtained through collateral assets and contributor assets. In France, the Monge ship, which has five radars including ARMOR, not only is intended to collect information on missiles in flight but also can occasionally provide very precise monitoring of orbital elements in LEO. In the UK, the Chilbolton Facility for Atmospheric and Radio Research is under the authority of the Rutherford Appleton Laboratory. Although mainly dedicated to atmospheric and ionospheric research, due to a recent update, the CAMRa[9] may contribute to the characterisation of objects in orbit. The Starbrook and Starbrook north wide-field telescope located at RAF Troodos in Cyprus and funded by the BNSC can be used to classify objects in GEO.

Similarly, the Scandinavian system EISCAT (European Incoherent Scatter Scientific Association) is intended to study the ionosphere and comprises a monostatic VHF radar in Tromsø, with two reception sites in Sweden (Kiruna) and Finland (Sodankylä). It can be partially used for purposes of monitoring in LEO, especially for polar orbits (EISCAT, 2022).

Optical means can also be used; for instance, the French systems SPOC (Système Probatoire d'Observation du Ciel), ROSETTE and TAROT (Télescope à Action Rapide pour les Objets Transitoires) as well as Britain's PIMS (Passive Imaging Metric Sensor) using a telescope at Herstmonceux in the UK, another in Gibraltar and a third one in Cyprus. Other examples include the astrometric telescope Zimlat in Switzerland, whose main mission is laser telemetry and which can be used to watch objects in GEO and produce accurate trajectography, and the ZIM SMART telescope at the University of Bern, which is used to establish a catalogue of orbital elements in GEO, GTO and MEO and allows to identify small objects. Finally, the Italian systems, Croce del North and the ASI's multistatic radar system, were designed mainly for the observation and detection of debris and NEOs, as well as the TFRM telescopes at the Montsec Astronomical Observatory and the Sagra Sky of the Observatory of Mallorca whose main missions are the same. Among these building blocks, Norwegian radar Globus II and British radar BMEWS have a specific status as they are part of the American SSN monitoring system and data are not available on an open basis.

As far as the SST segment is concerned, ESA relies on these national capabilities with a special focus on building up a coherent model, one taking into account the very sensitive issue of information quality and data policy.

Conclusion: What Is at Stake for Europe?

Today, Europe faces a number of necessary decisions. Space surveillance is now one of the concerns of all space actors. Questions of security and safety have become unavoidable. China is gradually forming a ground network and establishing its catalogue. Japan relies on its cooperation with the United States in attempting to enhance the American facilities settled on its territory by establishing its own sensors. India is also in the process of acquiring autonomous means in relation to its priorities and its technical and financial resources. At the end of the day, alongside public parties, private parties are also emerging mainly in the field of traffic management.[10]

Europe has significant capabilities, which it has to enhance in order to improve its competences and better address its own needs, as well as to reinforce its international position. The sharing of national capabilities still needs to overcome obstacles related to the specific aspect of intelligence data and more broadly to security concerns in bilateral and multilateral exchanges. These obstacles go beyond the question of space surveillance. As long as these needs do not seem to be a priority, they hinder the allocation of a significant budget because of a lack of a clear policy about data transparency. However, it can be noted that the space industry is increasingly keen to promote technical solutions for surveillance and cleaning up debris regardless of their origin, including those caused by ASAT tests.

The question of space surveillance has also reached a new diplomatic dimension. Since 2014, the United States has implemented a Stratcom SSA Sharing Strategy aiming to better disseminate and therefore better handle the data coming from the American network. In parallel, at the Vienna meeting in June 2016, Russia suggested to the UN Committee on the Peaceful Uses of Outer Space to create a UN-run database "collecting, systemising, sharing and analysing information on objects and events in outer space". This written proposal marks a turning point in the Russian attitude as it offers to make public its own catalogue on near-Earth objects, including military satellites not covered by the open catalogue of the North-American warning system NORAD, that the catalogue of the US and their Allies. The disclosure of this sensitive data raises many questions in terms of data policy. From the Russian point of view, an international database would help transparency as the relevant data would be available to any country possessing space assets. The sharing of government data banks would contribute to the openness of information and lay the foundations for true cooperation. At the same time, such a proposal undermines the monopoly of the US's unique position in regulating outer space traffic and represents a potential threat to US military satellites. However, the fact is that the NORAD system has always included Russian, Chinese and other space faring military satellites except in when there was a specific bilateral agreement.

The issue of transparency is not new. This kind of proposal promoting transparency as a key incentive for better stability has a very long history: the ISMA (International Space Monitoring Agency) and PAXSAT proposals date back to 1978 and 1986. Interestingly, they were proposed by France and Canada at that time but simply ignored by the US and the Soviet Union which operated their own systems.

Today, the increasing number of catalogues of space objects represents a new opportunity to balance the balance of power. SSA cooperation has given Europe a true experience of the challenges related to cooperation between different kinds of players on a wide range of issues, from debris to defence-oriented objects and data policy. The fact is that today's tense international situation no longer allows us to hope for increased transparency on a basis of mutual trust.

The fact is that the international situation is now so tense that there is no longer any hope of increased transparency on a basis of mutual trust. The Russian proposal has not been implemented and is unlikely to be implemented in the near future. Moreover, space surveillance is an increasingly sensitive issue with tens of thousands of satellites in constellations occupying low orbits, thus increasing the risk of collisions. Europe's efforts are now part of a broader perspective, that of Space Traffic Management, which must take into account the initiatives of private players in parallel with those of government entities to ensure the safety of its resources in orbit (European Commission, undated). At this stage, there is a true opportunity for Europe to play a more prominent role at the international level (Sourbès-Verger, 2016: 3–19). The ability to master situational awareness, from debris tracking to monitor hybrid threats to space infrastructures, is surely a strategic concern in terms of technological and political non-dependence.

Notes

1 The altitude at which these debris orbit as well as their size have a major impact on their lifespan. The larger ones, and those whose perigee is the lowest, re-enter the atmosphere under the gradual effect of gravity and naturally burn during the process. According to experts, 700,000 objects larger than 1 cm and 170 million objects larger than 1mm are expected to reside in Earth orbits.
2 Center of Control of Outer Space.
3 Diamant in the case of France and Black Arrow in the case of the United Kingdom.
4 In the 70s, his findings are quoted as open source by Charles S. Sheldon II, chief of the Science Policy Research Division of the Library of Congress, author of a very well-known annual publication on Soviet Space Program, published by the Congressional Research Service, Library of Congress, USA.
5 See the 32 papers published in the framework of the History Study Reports covering the period 1959–2003 (Noordwijk: ESA Publications Division ESTEC). https://www.esa.int/About_Us/Corporate_news/History_Study_Reports.
6 See http://www.orbitaldebris.jsc.nasa.gov and http://www.celestrak.com/satcat/boxscore.asp. Detected debris are larger, from 10 cm in LEO to 1 metre in geostationary orbit.
7 Kontrolya Kosmicheskogo Prostranstva in Russian, formely known as TsKKP.
8 Limited to detection at a local level (azimuth 180°).
9 A 25-metre steerable parabolic dish, for a meteorological S-band radar.
10 The Space Data Association (SDA), for example, tracks objects in GEO orbit in order to prevent collisions, avoid interference and geolocate the sources of harmful interference. Members of the SDA are satellite operators – both public and private – including EUMETSAT and Eutelsat at the European level.

Bibliography

Al-Rodhan & Nayef, R. F. (2012), *Meta-Geopolitics of Outer Space. An Analysis of Space Power, Security and Governance*, London: Palgrave Macmillan.

ARPA. (1959), 'Progress Report on Military Space Project', Quarter ended 31 December 1959, p. 27. Available at: http://www.nro.gov/foia/declass/WS117L_Records/471.PDF

Brachet, Gérard. (2016), 'The Security of Space Activities', *Non-Proliferation Paper* no. 51, Stockholm: SIPRI, July 2016, pp. 2–9. Available at: https://www.sipri.org/publications/2016/eu-non-proliferation-papers/security-space-activities

Caleb, Henry. (2015), 'Five European Countries Sign Space Surveillance and Tracking Agreement', *Via Satellite*, 17 June 2015. Available at: https://www.satellitetoday.com/government-military/2015/06/17/five-european-countries-sign-space-surveillance-and-tracking-agreement/

Committee for Civilian Crisis Management. (2006), 'Outline of Generic Space Systems Needs for Civilian Crisis Management Operations', 26 June 2006.

Council of the European Union. (2003), *A Secure Europe in a Better World, European Security Strategy*, Brussels, 12 December 2003.

Council of the European Union. (2014), Decision No 541/2014/EU Establishing a Framework for Space Surveillance and Tracking Support, 16 April 2014. Available at: https://
eur-lex.europa.eu/legal-content/EN/TXT/PDF/?uri=CELEX:32014D0541&from=EN

Davis, S. (2016), 'The Story of the Kettering Radio Group', *National Space Centre*, 31
August 2016. Available at: https://spacecentre.co.uk/blog-post/kettering-radio-group/

Decision of Council Joint Action 2001/555/CFSP Available at: http://www.consilium.europa.
eu/uedocs/cmsUpload/l_20020010725en00050011.pdf repealed by Council Decision
2014/401/CFSP of 26 June 2014. http://publications.europa.eu/en/publication-detail/-/
publication/c335c2a5-fdcd-11e3-831f-01aa75ed71a1/language-en.

del Monte, Lucas. (2015), *The ESA's Space Situational Awareness Initiative: Contributing to a Safer Europe*, pp. 1–6. Available at: https://www.researchgate.net/
publication/265894241_The_ESA's_Space_Situational_Awareness_initiative_contribut-
ing_to_a_safer_EuropeEC/

ESA. (2010), *EC/ESA Joint Secretariat Paper on Space and Security*, February 2010.
Available at: http://www.europarl.europa.eu/meetdocs/2009_2014/documents/sede/dv/
sede170310ecesaspaceandsecurity_/sede170310ecesaspaceandsecurity_en.pdf

EISCAT. (2022), 'Welcome to EISCAT Scientific Association', last updated: 21 November
2022. Available at: https://eiscat.se/about

ESA Council. (2000), 'Resolution for a European Policy on Protection of the Space
Environment from Debris', 20 December 2000.

European Commission. (undated), 'An EU Approach for Space Traffic Management:
For a *Safe, Secure* and *Sustainable Use* of Space'. Available at: https://defence-
industry-space.ec.europa.eu/eu-space-policy/eu-space-programme/eu-approach-space-
traffic-management_en

European Commission. (1988), T*he Community and Space: A Coherent Approach*, COM(88)
417 final, 26 July 1988 p. 24. Available at: http://aei.pitt.edu/3821/.

European Commission. (2000), 'Europe and Space: Turning to a New Chapter', *Communication from the Commission to the Council and European Parliament*, COM(2000) 597
final, 27 September 2000. Available at: https://eur-lex.europa.eu/legal-content/EN/TXT/
PDF/?uri=CELEX:52000DC0597&from=EN

European Commission. (2004), 'On the Implementation of the Preparatory Action on the
Enhancement of the European Industrial Potential in the Field of Security Research –
Towards a Programme to Advance European Security Through Research and Technology',
Commission Communication, COM(2004) 72 final, 3 February 2004.

European Commission. (2021), *EU Space Surveillance and Tracking: Service Portfolio*.
Available at: https://www.satcen.europa.eu/keydocuments/EUSST_Service_Portfolio622b
4b653172450001cfa321.pdf

European Parliament. (2008), *Resolution on Space and Security*, EP document 2008/
2030(INI). 10 July 2008. Available at: http://www.europarl.europa.eu/sides/getDoc.
do?pubRef=-//EP//TEXT+TA+P6-TA-2008-0365+0+DOC+XML+V0//EN.

European Space Agency. (1988), *Space Debris. The Report of the ESA Space Debris Working Group*, ESA SP-1109, 1988.

Flohrer, Tim, Krag, Holger, Klinkrad, Heiner & Schildknecht, Thomas. (2011), "Feasibility
of Performing Space Surveillance Tasks with a Proposed Space-Based Optical Architecture", *Advances in Space Research*, volume 47, Issue 6, pp. 1029–1042. Abstract;
Available at: http://adsabs.harvard.edu/abs/2011AdSpR..47.1029F

Joseph, Robert G. (2006), *Remarks on the President's National Space Policy – Assuring
America's Vital Interests Under Secretary for Arms Control and International Security*,
Washington, DC: The George C. Marshall Institute. Available at: https://2001-2009.state.
gov/t/us/rm/77799.htm

Klinkrad, Heiner. (2006), *Space Debris, Models and Risk Analysis*, Chichester: Praxis Publishing.

Klinkrad, Heiner. (2008), 'Europe's Eyes on the Skies. The Proposal for a European Space Surveillance System', *ESA Bulletin* no.133- February 2008, pp. 42–48. Available at: http://www.esa.int/esapub/bulletin/bulletin133/bul133f_klinkrad.pdf.

Krige, John & Russo, Arturo. (eds) (2000), *A History of the European Space Agency 1958–1987, vol.1- ESRO and ELDO,1958–1973*, Noordwijk: ESA Publications Division, pp. 263–271. Available at: https://www.esa.int/esapub/sp/sp1235/sp1235v1web.pdf

Lele, Ajay. (ed) (2012), *Decoding the International Code of Conduct*, New Delhi: Pentagon Press. Available at: https://www.academia.edu/20671878/Decoding_the_International_Code_of_Conduct_for_Outer_Space_Activities

Lele, Ajay, Gupta, Arvind & Mallik, Amitav. (2012), *Space Security: Need for Global Convergence*, New Delhi: IDSA.

Pasco, Xavier. (2009), *A European Approach to Space Security*, Cambridge, MA: American Academy of Arts and Sciences, pp. 6–7. Available at: https://www.amacad.org/publications/spaceEurope.pdf

Pellegrino, Massimo & Stang, Gerald. (2016), *Space Security for Europe*, Report - no.29, ISS European Union Institute for Security Studies, pp. 23–29. Available at: http://www.iss.europa.eu/publications/detail/article/space-security-for-europe

Pelton, Joseph N. & Jakhu, Ram. (2010), *Space Safety Regulations and Standards*, IAASS Book Series, Oxford: Elsevier.

Schrogl, Kai-Uwe, L. Hays, Peter, Robinson, Jana, Moura, Denis & Giannopapa, Christina. (2014), *Handbook of Space Security: Policies, Applications and Programs*, New-York: Springer-Verlag.

Sourbès-Verger, Isabelle. (2016), *EU-India Cooperation on Space and Security*, Roma: IAI Working Papers, 2016, pp. 3–19. Available at: https://www.iai.it/sites/default/files/iaiwp1638.pdf

Space and Security Panel of Experts. (2005), 'Report of the Panel of Experts on Space and Security', Available at: https://www.dlr.de/rd/Portaldata/28/Resources/dokumente/rp6/Bericht-SPASEC.pdf

Space Security Index 2016, 13th edition (2016), Waterloo, Ontario: Project Ploughshares. Available at: www.spacesecurityindex.org

Stares, Paul. (1985), *The Militarization of Space, US Policy 1945–1984*, Ithaca, NY: Cornell University Press, pp. 131–133.

Technological and Aerospace Committee of the WEU. (2007), *Weapons in Space Report*, rapporteur: A. Meale, 21 June 2006.

Treaty of Nice (2001), Official Journal of the European Communities C 080, 10 March 2001, pp. 1–87.

Treaty of Lisbon (2007), Official Journal of the European Communities C 306, 17 December 2007, pp. 1–271.

UNIDIR. (2006), *Safeguarding Space Security: Prevention of an Arms Race in Outer Space. Conference Report 21–22 March 2005*, Geneva: UNIDIR.

United Nations Office for Outer Space Affairs. (undated), 'Online Index of Objects Launched into Outer Space'. Available at: http://www.unoosa.org/oosa/osoindex/search-ng.jspx?lf_id=

United Nations Office for Outer Space Affairs. (2004), 'European Code of Conduct for Space Debris Mitigation', Issue 1.0, 28 June 2004, Available at: https://www.unoosa.org/documents/pdf/spacelaw/sd/2004-B5-10.pdf

Verger, Fernand, Sourbès, Verger Isabelle & Ghirardi, Raymond. (2003), *The Cambridge Encyclopedia of Space*, Cambridge, MA: Cambridge University Press, pp. 341–343.

6   The European Space Industry as a Driving Force for Militarization

Iraklis Oikonomou

Introduction

With the adoption of the 2021–2027 EU Space Programme, it is now safe to admit that there has been an irreversible drive towards the inclusion of security and defence into EU space policy, i.e. the opening up of that policy to military applications, needs, and uses. New flagship projects, like Governmental Satellite Communications (GOVSATCOM), the Secure Connectivity Initiative, and Space Traffic Management, constitute the second wave of militarization of EU space, expanding and deepening an already existing phenomenon that started with the security and defence applications of Copernicus and the Public Regulated Service (PRS) of Galileo. Satellite navigation, earth observation, satellite communications, and space situational awareness – the Union seems to have at last managed to establish the foundations for a full merging of civilian and military applications into the portfolio of programmes that comprise its space programme.

When dissecting the emergence of this defence dimension of space policy, industrial actors are usually depicted as merely those who are supposed to produce what states and European institutions ask for in terms of capabilities. In other words, their political role as drivers of militarization is not appreciated fully – crushed among the supranational Commission, the intergovernmental European Space Agency (ESA) and the myriads of member-state interests. This chapter will introduce industrial interests as a key source of policy transformation, examining in particular the discourse and positions set by the space industry and its Brussels lobbying organization, ASD-Eurospace, and the extent to which the outcome, i.e. the emergence and solidification of EU military space, is compatible with industrial preferences and stems from the industrial vision.

Adopting a historical materialist perspective, the chapter is an attempt to establish the crucial part played by internationalized space-industrial capital in promoting military space agenda in the EU. This trend will be documented via material deriving from official discourse and publications by the industry. Also, it will be interpreted by examining key aspects of the global political economy of space production that fuel the industry's quest for more security and defence space programmes at the Brussels level. Overall, it will be argued that the European space manufacturers facilitated the emergence of a non-civilian

DOI: 10.4324/9781003230670-9

dimension in European space activity, seeing in it a huge opportunity for market expansion and competitive survival.

Theoretical and Empirical Aspects of EU Space-Industrial Actorness

The starting point of the present analysis is the definition of the process of European integration as 'very much the process by which European society has been transformed to allow the imposition of the discipline of capital on a scale beyond the national state' (Holman & van der Pijl, 2003: 79). However, certain modifications are necessary: First, we are not dealing with European integration in general, but rather with European space policy integration, and thus, the emphasis is on the discipline of a specific fraction of capital, hereby termed space-industrial capital. Second, the transformation does not concern primarily the social (European society) but rather the politico-institutional realm, given that the object of analysis is programmes established by EU institutions. Third, the idea of militarization denotes a specific kind of integration, based on the promotion of a non-civilian dimension of space policy. Therefore, what needs to be explored is not simply the integrative process generated by the imposition of the discipline of capital versus the lack of integration; it is the production of a specific essence of integration (promotion of a militarized policy) versus another (promotion of a civilian policy).

From this general framework stems a crucial theoretical principle: the idea of space-industrial capital as a social force (Cox, 1996: 100–101; Holman & van der Pijl, 2003: 72–73). In other words, the companies involved in space production are not seen here as just economic units of analysis; they share particular visions of the world and of what the EU should be and should do and proceed with the shaping of relevant policies at the Brussels level via their own political representation organizations. Historical materialist thinking pays particular attention to this political actorness of capital, with Gill and Law (1993: 95) noting business' 'privileged ability to influence governments' by mobilizing direct and structural forms of power. The former involve a wealth of resources, contacts within the government, control of the media establishment, expert knowledge, etc. The latter implies a deeper dimension of power that stems from the position of capital within the production system, dominating the market, employment, investment, innovation and growth (Gill & Law, 1993: 99–100), as well as what constitutes the discursive articulation of the general interest of the society as a whole. With this theoretical predisposition in mind, the space industry is expected to have set the conceptual, discursive, ideational terms of the debate on the need to militarize EU space policy – which is the main hypothesis of this chapter.

The primary fraction of European space-industrial capital comprises the two large system integrators: Airbus Defence and Space, and Thales Alenia Space. The two have been the outcome of a long process of consolidation, whereby smaller-scale national champions internationalized, establishing a long web of international business interconnections through a wave of domestic and cross-border mergers and acquisitions. But not all system integrators maintain an internationalized structure – OHB, for instance, has had a primarily German origin and national champion status, ever since its establishment as a space company in the 1980s. Next to system integrators, there are sub-system suppliers and equipment suppliers, comprising the landscape of European space companies characterized by varying portfolios and sizes.

Thales Alenia Space is essentially a French-Italian entity, jointly owned by Italian defence firm Leonardo (formerly Finmeccanica) and the French defence firm Thales Group. Its roots are to be found in the acquisition, by Thales Group, of Alcatel's share in Alcatel Alenia Space and Telespazio, two joint ventures between the French telecommunication and electronics firm and Finmeccanica. Alcatel Alenia Space was established in 2005 through the merger of Alcatel Space and Alenia Spazio and was co-owned by Alcatel and Finmeccanica. Approximately a year later, in April 2006, Alcatel sold its participation in Alcatel Alenia Space and Telespazio to Thales Group.

Airbus Defence and Space is the outcome of a similar consolidation path, involving first and foremost the creation of EADS through the merging of German DASA, French Aerospatiale-Matra and Spanish CASA in 1999. In parallel, the merger of the space divisions of the three companies led to the formation of Astrium in 2000. In 2003, EADS became the sole owner of Astrium through the acquisition of BAE Systems' 25% participation in it. Ten years later, Airbus Defence and Space incorporated Astrium and became one of the three constituent entities of Airbus Group, together with Airbus and Airbus Helicopters, in the context of the broader reorganization of EADS into the Airbus Group.

The main space contractors have established their own interest representation group, ASD-Eurospace. The organization was established in 1961 as Eurospace; in 2004, it became the dedicated space group of the Aerospace and Defence Industries Association of Europe. The members of ASD-Eurospace are European companies active in the production of space systems and the main aim of the organization is the promotion of space activities in the interest of its members and the definition and expression of common views on behalf of European space manufacturers (ASD-Eurospace, 2022: slides 3, 7). Thus, it is critical to point out the dual function of the organization: on the one hand, the advocacy of industrial viewpoints at ESA and EU institutions and, on the other hand, the elaboration and consolidation of these viewpoints among the industrial community members. In other words, ASD-Eurospace mediates not only on behalf of the industry (vis-à-vis ESA, the Commission, the Parliament, etc.) but also within the industry, bridging gaps and conflicts that may arise based on sectoral divisions, geographical divisions, size of companies, etc.

Neither ASD-Eurospace nor individual space-industrial entities can be seen in isolation from their main structural feature: the merging of space-industrial and military-industrial capital under the same roof. To be more precise, this is not a merging on equal terms; space-industrial capital essentially falls under the broader umbrella of defence-industrial capital in at least two ways. At the industrial-economic level, the primary space contractors have been acquired by chiefly military conglomerates (Thales, Leonardo, Airbus). At the institutional level, the representation of space-industrial interests is connected to the representation of military-industrial ones, with the incorporation of Eurospace by ASD. The two levels are dialectically unified: no institutional merging in terms of lobbying would have been possible without the prior absorption of space companies by arms manufacturers through a set of international mergers and acquisitions.

Interaction between the European Commission and industrial lobbyists is profoundly dense through a number of formal and informal links. Indicatively, ASD-Eurospace is involved in the preparation of a European Co-programmed Partnership titled 'Global competitive space systems', to be included in the Horizon Europe framework. This preparatory process, stemming from the effort to set a strategic research and innovation agenda for EU-funded space research, involved a series of regular meetings and the exchange of input and information for a period of three years between the Commission and ASD-Eurospace. Apart from the latter, the consortium comprises the European research and technology infrastructure (European Association of Research and Technology Organisations), the European academic active in space research (European Aeronautics Science Network), the national space research centres (Association of European Space Research Establishments), and the representative organization of the SMEs of the European space industry (SME4SPACE) (Naujokaityte, 2021; ASD-Eurospace, 2022: slide 30). As for the rationale of such a collaboration? The summary of what the partnership is and does copy-pastes Europe's Space Strategy: 'Fostering a globally competitive and innovative European space sector' and 'Reinforcing Europe's autonomy in accessing and using space in a secure and safe environment' (DG DEFIS, 2020: 5).

But it is not sufficient to point out the deep and mutual interconnection between the Commission and ASD-Eurospace; the structural involvement in defence-related programmes also needs to be established. An indication of how involved industrial lobbyists are in the authoring of EU space militarization is the structure of ad hoc task forces within ASD-Eurospace. These task forces are mandated to tackle, in a coordinated manner, matters that arise in connection to European institutions and have a short-to-medium term horizon. If we exclude the long-standing REACH-related task forces, four out of seven existing task forces have a clear defence dimension. Specifically, these four deal with Space Traffic Management, Secure Connectivity, Copernicus NExtGen, and the European Defence Fund, with the three 'civilian' task forces being the ones on Horizon Europe, ESA Industrial Policy, and the European Launcher Alliance (ASD-Eurospace, 2022: slide 26).

The Phenomenon: Recent Militarizing Trends in European Space Policy

Until very recently, space was seen by the EU as an area where only civilian applications were allowed. True, the two existing flagship programmes, Galileo and Copernicus, contained security and defence-related segments, as has been shown elsewhere (Oikonomou, 2013, 2017). However, they were not purely military space applications, respecting – at least on the surface – the commitment of the ESA and the Union to civilian-only activities. Three new programmes, Government Satellite

Communications (GOVSATCOM), Union Secure Connectivity Programme, and Space Traffic Management, have altered this condition irrevocably; the EU now has ambitious military-space applications under way. Space manufacturers have been to a great extent part of the European arms industry, through a wave of mergers and acquisitions in the 1990s and the early 2000s. The availability of two additional military projects adds new opportunities for pan-European procurement, i.e. extra sales and profit for the industry.

As far as GOVSATCOM is concerned, the primary objective of the programme is the provision of secure satellite services for EU and member-state authorities. A 10 million-euro Preparatory Action is already up and running, paving the way for a full-fledged programme for the 2021–2027 period, through a range of projects involving satellite producers, end users, and satellite communication providers. Potential users include a variety of security and defence actors, such as the armed forces, border security forces, civil protection forces, and diplomatic services. As for the Union Secure Connectivity Programme, it lies conceptually and practically close to GOVSATCOM but is in fact a distinct programme, providing secure and reliable space-based broadband connectivity via multi-orbit satellite constellations. The programme is, indirectly, of a security and defence-related orientation too, expected to optimize 'effectiveness of surveillance, EU external action and crisis management activities' (Whittle et al., 2021: 53). Border surveillance, maritime surveillance and control, maritime emergencies and CSDP missions are all expected to benefit

The EU has engaged in the development of yet another military space programme for the establishment of a Space Traffic Management capability, with Space Surveillance and Tracking (SST) at its core. Following a Council Decision in 2014, an EU SST Support Framework was established and a consortium of eight states was set up (with the five founding states being also the primary space manufacturing ones – France, Germany, Italy, Spain, and the UK). The consortium eventually formed a cooperative scheme, called SST Cooperation, with the EU Satellite Centre. Three core actions fall under the scope of the Framework: establishment and operation of existing sensors to track space objects; processing of data; and service provision (Council of the EU, 2014: Art.4). In theory, SST is articulated as a programme with a purely civilian orientation – a response to the fact that 'due to the growing complexity of the orbital environment, space-based assets are increasingly at risk from collision with other operational spacecraft or debris' (EUSST, 2020: 4). However, there is also a military dimension to SST, no matter how hard the EU institutions have tried to conceal it. True, SST concerns the defence of space and, thus, deviates from all other EU space projects that seek to utilize space for defence (Polkowska, 2020: 128). But this does not make it less military-oriented or more civilian. The detection and identification of man-made objects include objects used by other states for military purposes. In other words, if space is a potential battleground, its monitoring is evidently a military mission and purpose.

Until now, no direct benefit for the space industry can be traced to the efforts of SST Cooperation, mainly because the initiative involves the coordination of existing resources rather than the generation of new ones. Nevertheless, if considered in a dynamic context, the EU has opened the door to an immense set of opportunities for space manufacturers, given that the option of developing additional capabilities in the future cannot be excluded. Under the coordination of Italy, and with the participation of France, Germany and the Netherlands, the European Military Space Surveillance Awareness Network is now a Permanent Structured Cooperation project. Its goal is to develop an autonomous EU military capability for space situational awareness, to be integrated with the EU SST Framework. Thus, the addition of new capabilities is a real possibility, translating into new investment, sales, and profits for the industry.

In February 2022, the European Commission together with the High Representative for Foreign Affairs and Security Policy released a joint communication proposing the main elements of an EU approach to space traffic management. From the outset, space traffic management is placed in a defence context, as it is expected to 'contribute to the security and defence dimensions of the EU in space' with space assets 'becoming targets of various kinds of threats' (European Commission & High Representative, 2022: 1). Initially, the façade of the Commission's logic appears technocratic, with references to the increase in the number of satellites and the consequent risk of collisions. However, soon this is replaced by a much grander perspective that, as we shall see, reflects the industry's positioning on the matter, with a call to 'increase EU resilience by avoiding technological dependencies' and 'ensure strategic autonomy through the development of EU capacities' (European Commission & High Representative, 2022: 6). As for the 'stakeholders', it is easy to guess the one that is explicitly mentioned following this excerpt: the EU space industry, whose competitiveness the programme is expected to support with additional funds and the creation of a relevant European ecosystem around SST technologies and capabilities (European Commission & High Representative, 2022: 8).

In addition, the Commission has declared its intention to have the defence dimension of EU space policy further strengthened. With its February 2022 Communication on the Commission's contribution to European defence, it opened up space as a field of military applications to an extent that just some years ago would have been unthinkable. The list of actions is endless, starting from the further protection of EU space assets through SST, Space Situational Awareness, and early warning capabilities for defence. A second pillar involves the upgrading of defence features in existing EU space infrastructure, moving beyond the PRS of Galileo, with the inclusion of defence requirements in Copernicus and the establishment of the Union Secure Connectivity Programme based on GOVSATCOM. In this context, the EDF is expected to fund military space programmes, like the Galileo for EU Defence, space-based navigation warfare, and military space surveillance awareness. Third, the Commission has committed to working towards the reduction of space-related EU strategic dependencies on critical technologies by mobilizing the full range of its space-related initiatives, from EDF to the Space Programme. And, finally, as far as governance is concerned, the Commission plans to introduce a Galileo PRS-like model in the all elements of EU space infrastructure and in Copernicus in particular, with a dedicated governmental service that will be taking into account defence requirements (European Commission, 2022: 10–13).

Next to the Commission, the European Defence Agency (EDA) has been equally active in promoting the military dimension of EU space with a range of activities and initiatives. For instance, on earth observation, the EDA has been compiling a collection of high-level user requirements for future military satellite constellations, while also investigating how Copernicus could be modified in order to better support military operations. As far as satellite navigation is concerned, the Agency has provided support towards the development of a European military satellite navigation policy, with the endorsement of a Common Staff Target, which should lead to an improved usability of Galileo for defence users (EDA, 2018b). Above all, one cannot ignore the key role that EDA has played in GOVSATCOM. Specifically, the Agency initiated the process of the programme's development, with the identification of military user needs and requirements and proceeded with the setting up of an EDA GOVSATCOM demonstration project. The goal of the project has been to prove the concept and demonstrate the benefits of pooling the satellite communication capabilities of contributing member states (EDA, 2019). Last but not least, the EDA has established the EU Satellite Communications Market, pooling the demand for commercial satellite communication and wider communication and information services, with military SatCom being part of the offered range of services (EDA, 2018a).

The Discourse: Strategic Autonomy – From the Industry, for the Industry

The discursive call for more defence in EU space has consistently and constantly originated from the industry. Of course, for this call to succeed, it has to appear in isolation from the interests of the industry for more sales and profits, serving instead a much broader goal. Sometimes, this goal is economic growth or the improvement of public policies; take, for instance, industry executive Jean-Loic Galle's (in ASD, 2016: 1) address to the EU to promote 'a better and increased use of the – insufficiently tapped – capacities of space to create growth and to support and improve public policies (in particular in the areas where space can contribute to achieving the sectoral objectives set out by the EU in the fields of environment, transport, security & defence or digital economy)'.

But most often, the conceptual tool utilized by the industry in its effort for a more military-oriented EU space policy is strategic autonomy. In fact, this well-known and widely advertised notion has been inspired intensely by industrial vision and discourse. In 2007, several years before its appearance in Brussels' narrative, ASD's Director of Space at the time welcomed the Commission's decision to finance Galileo, claiming that it paved 'the way for the autonomy of Europe, regarding positioning and navigation by satellite' (Jean-Jacques Tortora, quoted in ASD, 2007). Autonomy was a recurring theme in industrial discourse prior to its emergence as the core and overarching concept of EU security and defence policy: 'To reap all benefits of space activity, Europe must ensure that it maintains alive the industrial capabilities to undertake space programmes with the appropriate level of autonomy' (ASD, 2009: 2). The conceptual chain follows a specific pattern: autonomy → capabilities → investment. Such a pattern can be traced, indicatively, in the following recommendation addressed from industry's lobbyists to the Commission:

> A priority for the next European Commission should be to decline [sic] operationally, in due cooperation with the Member States, the ambitions expressed in the pillar of the Space strategy calling to "reinforce Europe's autonomy in accessing and using space in a secure and safe environment". This would require to address the missing capabilities Europe needs to be equipped with to ensure its awareness, autonomy and freedom of action (i.e. security of EU-owned infrastructures in space and security from space).
>
> (ASD-Eurospace, 2019: 2)

The same notion appears through the term 'independence'; as early as 2009, the lobbyists of space manufacturers in Brussels reminded the European public of the fact that 'space has become critical for the independence that Europe should aim at' (ASD, 2009: 1), while a decade later, the then president of ASD-Eurospace stressed that 'space systems and related European access to space represent crucial elements for the independence and sovereignty of our continent' (Galle, 2018). Independence is also closely associated with non-dependence. In the technological domain, independence implies the development of all required space technologies in Europe, while non-dependence involves Europe's ability to secure unrestricted access to any space technology it may require (European Commission, 2021b: 4–5). In practice, the two terms are used interchangeably and have been adopted with equal fervour by the space manufacturers: 'Europe's non-dependence in space needs to be guaranteed; this implies the capacity to conceive, develop, launch, operate and exploit cost-effectively space systems, but also the necessity to rely on an unrestricted access to state-of-the-art technology' (Jean-Loic Galle in ASD, 2016: 1). It should be noted that such a vocal reiteration of the need for non-dependence came not from a European policy official or parliamentarian, but rather from the then president of ASD-Eurospace and former CEO of Thales Alenia Space. And certainly this is not a one-off case; 'enhancing European technological non-dependence in the space sector' is one of the key pillars of the argumentation by the lobbyists of space companies in support of a European industrial policy for space (ASD-Eurospace, 2017: 1).

Why should the industry be so vehement in its pursuit of European autonomy? Because beyond its strategic essence, autonomy constitutes a technological and industrial context that is greatly favourable to space manufacturers: put simply, 'ensuring autonomy is crucial ... for future industrial competitiveness' (Whittle et al., 2021: 13). Full-scale protectionism is the logical outcome of technological non-dependence. In fact, the latter has become the starting point of an entire list of industrial demands that would have been unthinkable without the security and defence umbrella. Their common feature is the elimination of non-European competition.

The industry has, among others, requested the restriction of competition when essential EU interests are at stake, the granting to the industry of the right of first refusal concerning non-European competitors, the inclusion of reciprocity in internationally open procurements, and the assessment of the exposure of European institutions to 'unplanned investment in order to secure industrial capacities and/or to restore hampered industrial competitive standing originally built up over decades of public investment', when a contract is awarded to a non-European competitor (ASD-Eurospace, 2017: 2).

Yet another concept put forward by the industry is 'technological sovereignty', which 'should be considered as an objective for those technologies that make a decisive contribution to a capability that is key for a critical function of a strategic sector' (ASD, 2020: 5). For the industry, technological sovereignty involves a degree of self-sufficiency, which is achievable through the existence of EU-based suppliers that command the relevant technology and the capacity to turn it into applications – a process that must be under European control for sovereignty to exist. Essentially, technological sovereignty is a synonym for technological independence, i.e. the avoidance of 'dependencies that would enable a non-European actor to unilaterally impose constraints on European technologies' (ASD, 2020: 5). Under conditions of technological sovereignty, EU-based producers are expected to face no obstacles from foreign industrial actors in completing successfully all stages of technology and product development.

What about satellite communications? In its 'full support' for the GOVSATCOM initiative, the industry's lobbying group, ASD, referred to the EU member-states' 'ability to respond autonomously' as the key rationale of such a programme. Specifically, 'a shared European capacity in satellite communications would significantly contribute to the EU's autonomy in decision-making and action at the global level in response to challenges relating to defence, security, humanitarian crises and natural/emergency disasters' (ASD-Eurospace, 2014: 3). Autonomy becomes an industrial demand, as it is – in the eyes of the industry – the precondition for safeguarding industrial competitiveness. Or, to be more precise, the quest for autonomy generates the political context through which the more industrial and economic demands can be fulfilled. A sizeable budget, a quick-paced process, and the involvement of the industry in it can only be demanded because they eventually serve the biggest, 'objective' goal of tackling a defence-related critical capability shortfall. Metaphorically, the industry sells 'autonomy' and buys the maintenance of a 'leading role on the export market and the competitiveness of European space industry' (ASD-Eurospace, 2014: 4).

The picture is not different as far as SST is concerned. In 2016, ASD set the following goal for the EU: 'Europe shall establish capabilities for independent assessment of the orbital environment (debris and objects tracking, and trajectory prediction) and support technological readiness for debris mitigation and prevention' (ASD-Eurospace, 2016: 20). Crucially, this was not some kind of prediction or empirical guess – it was put forward as a recommendation. And the proposal by the industry was comprehensive, detailing the specific needs of the ground and space segments, including tracking systems and sensors.

Overall, the industry has over and again emphasized how crucial it is for the competitive health of European space manufacturers to ensure the merging of space and EU ambitions in security and defence. Take the European Defence Fund and how warmly it was received by the lobbyists of the space industry in Brussels: its establishment 'can therefore offer a new opportunity to boost institutional investment in strategic and military applications of space – the central pillar of American, Russian and Chinese space policies'. Or take the grievances expressed by industrialists over the 'limited space military programmes in Europe', a reality that supposedly distorts competition in favour of non-EU industrial actors and at the detriment of EU ones. The common denominator: security and defence 'have a key role to play to foster the competitiveness and innovativeness of the EU's space technological and industrial base' (all excerpts from ASD-Eurospace, 2019: 2).

CSDP features prominently in the industry's discursive depiction of the necessity of space. In its contribution to the EU 2020 Strategy, ASD noted that 'space systems are strategic assets, facilitating the development of an autonomous European decision process for the benefit of all Member States and contributing to the efficiency of the ESDP operations' (ASD, 2009: 1). But gradually CSDP proved too little and too restrictive to support the grand vision of industrialists: the removal of the civil/military division in European space policy via the unification of the two realms.

There is regularly a debate in Europe on the military or civil character of these policies and of the infrastructures, organizations and industries which support these policies. Let me point out that, in the USA, Russia, China and India, this debate would be pointless. The US NOAA (weather forecasting), NIMA (mapping), NASA (Space), the GPS program, the space imagery programs, the secure communication networks between embassies have both civil and military justification but the public policies they support are civil AND military (Cipriano, 2007: 2).

And this, one might argue, is precisely the direction pursued by the second wave of the Union's military space programmes.

The Material Foundation: The Political Economy of European Space Militarization

The industry itself admits that 'today, after decades of consistent public investment, European industry has achieved enviable positions on the global market. Space is one of the areas of European technological excellence' (ASD-Eurospace, 2016: 2). However, past achievements are not enough as the global space landscape is characterized by profound change and transformation with the emergence of new competitive actors: 'The growing competition from China, and the new space business models pioneered in the USA require specific attention. The European Space Industry is competitive, but competitiveness needs to be permanently re-assessed with respect to the achievements of the competition' (ASD-Eurospace, 2016: 2)

To understand the political economy of space militarization, it is first essential to appreciate the importance of the European institutional markets, i.e. of public demand, for the European space manufacturers. Public demand is the cornerstone of predictability and stability of activities of space industries in general and the European space industry in particular. And, according to the latter, adding a strong layer of security and defence in space has the potential of guaranteeing 'a high and continuous level of public investment and a secured domestic market essential for ...competiveness' (Galle, 2018: 1). The admittance by former head of ASD-Eurospace, Giuseppe Morsillo (2016), is striking in its simplicity and bluntness: 'When benchmarking Europe with other main space-faring nations, clear structural weaknesses can be highlighted on the European side: indeed, there are limited space military programmes as compared to all other space powers (USA, Russia, China)'.

Indeed, the comparison with other space powers is a constant feature of the pro-industrial argumentation favouring the militarization of EU space. The Commission has been at the forefront of this trend, with the following statement being its guiding rationale:

> The European space industry differs from its main international competitors to the extent that its budget is smaller, it relies more on commercial sales, the share of military expenses is smaller and synergies between civil and defence sectors are far less developed.
>
> (European Commission, 2013: 5)

This Commission observation echoes a constant concern by the industry: 'In Europe, differently than in the USA, Russia or even China, military applications of space are rather under-developed' (ASD-Eurospace, 2020: 31).

The European institutional market can be conceptually split into sales of civil and military systems. During the period 1997–2006, the value of European civil programmes dropped considerably from approximately 2.4 billion euros to just above 2 billion euros. On the contrary, the military segment more or less doubled, from just below 500 million euros to almost 1 billion euros (ASD-EUROSPACE, 2007: 6). The military layer is a valuable addition to the industry's business as it translates into more revenue, and one that is of a more predictable and stable nature as it concerns public authorities only. Fluctuations in the civil market can be counter-balanced via orders for military systems, and that's why the absence of a generous military dimension has been dubbed a problem by the manufacturers: 'Strategic considerations have not been a major driver of space systems development in the early years of European space programmes, and today European space military programmes are still organized at national level rather than at European level' (ASD-Eurospace, 2020: 31).

This point is the key to grasping the industrial case for space militarization. The lack of military programmes has been tagged a problem by the industry itself, independently from the policymakers. In other words, we can trace a bottom-up process of uploading the industrial message to the policy elites, which is totally different from the idea that the industry is here to merely serve the needs and requirements set in a top-down fashion. The addition of a defence dimension has been expected to boost three interconnected domains: the European institutional investment in space, the size of the European domestic market, and the size of the R&D budgets. As de Concini and Toth (2019: 76) put it, 'the traditional European upstream space industry is used to a large institutional market of traditional public procurement and R&D grant programmes'. Military space is the means to strengthen what the industry expects and has already been used by the EU as a parallel form of public sector support.

Earth observation, satellite navigation, satellite communication, and space situational awareness all have dual-use applications. The military uses of space are ever-increasing, with the security domain being dubbed 'one of the most vibrant application domains in space' (de Concini & Toth, 2019: 31). But the specificity of security and defence as a service where the state is the sole customer means that the industry is heavily dependent on sales to public authorities, i.e. there is a very narrow customer base. In 2016, European institutional customers accounted more than half of the total business of the European space industry, out of which sales to military institutions were the third biggest category of revenue (573 million euros), after ESA (3.37 billion euros) and other civil public agencies (774 million euros) (de Concini & Toth, 2019: 84). This does not even convey the full significance of defence and security applications, as ESA is involved in both Copernicus and Galileo.

In fact, the rising trend of European space industry sales from the mid-2000s onwards seems to coincide with major developments in these two programmes, Copernicus and Galileo, and could be partly attributed to them. Specifically, in 2003, the EU took over the financing of Galileo following the collapse of the Public-Private Partnership, and in 2004, the Commission introduced the Action Plan that was to eventually lead to the establishment of Copernicus by 2008 – still called GMES at the time. In just over a decade, the annual final sales of the European space industry soared from just over 4 billion euros in 2003 to more than 8 billion euros in 2016 (ASD-Eurospace, 2021: slide 12). As for the contribution of EU programmes to total industry sales, this has been clearly documented on both fronts – earth observation and satellite navigation – with Copernicus' contribution skyrocketing from less than 100 million euros in 2010 to way over 400 million euros in 2017 and Galileo/EGNOS' contribution following a similar yet less impressive pattern, which can be easily explained by the earlier timing of the programme compared to GMES-Copernicus. The industry's own findings confirm the said trend: 'European Union programme's contribution to industry sales is increasing regularly since 2009' (ASD-Eurospace, 2020: 26).

Another benefit of militarization is the potential for synergies that they offer. Defence and security-related projects are advertised by the Commission as an opportunity for the facilitation of interactions between SMEs and established defence players. In parallel, the EU space-based global secure communications system is mentioned as a flagship project that could end up becoming a 'game-changer' for cross-fertilization 'due to its size or impact as well as its potential benefits for Europe's technological sovereignty' (European Commission, 2021a: 14, 16). In essence, the expectation here is to use militarization as a tool for boosting the innovation and competitiveness of the defence and space industry by attracting research organizations, SMEs, startups and other civilian industrial actors. Ideally, defence and space companies 'should be able to draw on EU civilian industry research achievements to avoid costly duplicated research' (European Commission, 2021a: 2).

All in all, space militarization is seen by analysts close to the industry as a tool – a driver, to be more precise – to achieve market development in the entire spectrum of space business. Due to the dual-use nature of much of the relevant technology and applications, i.e.:

> Due to the strong overlap of military, safety and security user needs, any system that serves one of these users will likely be able to prevail in the other sectors as the business conditions will be more favourable for such undertakings than for a total "outsider/newcomer", with no or limited exposure to the safety/ security/military requirements.
>
> (de Concini & Toth, 2019: 13)

This is an important point: with the gradual expansion of security and defence applications, EU military space becomes a field of business opportunities – a fertile ground where companies can find incentives to innovate and secure better conditions of competitive expansion in commercial markets too.

However, the adjective 'political' in political economy should not be underestimated or overshadowed by economic considerations. The defence 'turn' in EU space policy has made things easier for space manufacturers when it comes to promoting and legitimizing their demands. Space, in this respect, becomes something that exceeds our way of life and a set of (welcome) services – it becomes an existential precondition for survival. This sense of urgency is present in industrial discourse when it comes, for instance, to its demand for support and protectionism, under the label of technological autonomy and sovereignty. Defence and security are depicted as strategic sectors that require special treatment from the EU because they are essential for 'continuity of life' (ASD, 2020: 5); consequently, the inclusion of a defence dimension in space turns it, too, into such a sector where extraordinary measures can be applied in support of its competitive survival and expansion.

Elsewhere (Oikonomou, 2013, 2017), the politico-economic logic of both Copernicus and Galileo, parallel to their strategic value, has been detailed. GOVSATCOM confirms this pattern of adding value to the industry under the veil of a pressing strategic need and requirement. For instance, even though the Commission identifies the core problem to be the mismatch between the satellite communication needs of governments and what the Commission sees as appropriate solutions (European Commission, 2018: 14), the conceptual foundation of the initiative has been founded on the needs of the European space industry, 'especially in the context of strong international competition'. The Commission not only acknowledges that European space manufacturers are well-positioned globally, as they control 'one third of all global satellite sales', but also refers to the existence of 'other spacefaring nations' that 'have a much stronger and more stable domestic customer base, mainly in the form of national programmes' (European Commission, 2018: 20). What the Commission points to is the competitive threat of the US. However, the entire discursive base of legitimization is highly contradictory; why should Europe bother with what the US does if European manufacturers are already profitable and control a big chunk of the global market?

The picture becomes even more complicated the further the official narrative proceeds: 'European autonomy also requires a strong, innovative and globally competitive industrial base to design, build and operate the secure satcom systems, including space infrastructure, ground segments, network services, and user equipment' (European Commission, 2018: 30–31). Look at the circular argument evident here: Europe requires a strong space industry to manufacture the systems that are necessary for the space industry to remain strong! The argument also lacks logical coherence; it makes sense only as a legitimation effort on behalf of the Commission to justify the subsidizing of the European space companies. In other words, there is no objective need to maintain a globally competitive industry if the mission of that industry is to produce the systems that will safeguard its competitiveness. The only real need here is to facilitate the provision of additional EU funding to the hands of the manufacturers.

Conclusion

The main finding of this chapter is that the European space industry has not been a passive recipient of the Commission's drive to military space; instead, it has actively pressed for this process, providing critical input in the form of conceptual frames and vision. Furthermore, the present note established that the addition of a military layer in EU space policy contains a distinct socio-economic essence and function, far beyond the mainstream understanding of space as a tool for the fulfilment of security and defence purposes. What is being secured via EU military space is the profitability and sustainability of European space manufacturers, and what is being defended is the global competitive status of the industry vis-à-vis non-EU industrial actors. The mainstream logic of threats that have to be countered by developing capabilities, currently on the rise due to the Russian invasion of Ukraine, omits the mere fact that military space in the form of the first wave of militarization (Galileo PRS, Copernicus) emerged prior to any threats, and the same goes for the second wave too (GOVSATCOM, Secure Connectivity, Space Traffic Management, etc.).

True, contradictions persist and have characterized much of the evolution of EU military space. Not all member-states have a mature space sector and neither have all of them been equally keen to see the EU budget turn to defence purposes; or as an ASD-Eurospace lobbyist put it, 'it is a mixed bag, where everything is European, but at the same time the national interests stay strong' (Jérémy Hallakoun, quoted in Naujokaityte, 2021). Different state interests, mirroring the interests of the respective industrial capital fractions, translate into diverging visions, levels of ambition, and technological abilities. In addition, space is not the only sector in the European economy that requires state subsidies and support and the space. The industry has acknowledged this sectoral and national differentiation, as reflected in, among others, proposals to 'focus on technologies that are critical for a broad variety of applications from different strategic sectors and (potentially) used in many Member States' (ASD, 2020: 11). And, of course, bureaucratic divergences have persisted along the intergovernmental v. supranational spectrum, with the emergence of a new actor, EUSPA.

However, we are not anymore in the 1990s or the 2000s, when the nascent space identity of the EU was in the process of being established. Students of the EU have spent too many years lamenting the lack of initiative and capabilities, and pointing to the added value of a security and defence dimension of space that was supposedly nowhere to be found. Following the barrage of new military-related space programmes and the addition of a defence dimension to existing ones, the real question that the analysis of the EU must tackle is how and why the process of EU space policy militarization was successfully initiated, facilitated, and realized at the EU level. As this chapter highlighted, there can be no answer to this question without taking seriously into account the role of the European space-industrial capital as a politically organized and powerful social force.

Bibliography

Airbus. (undated), 'New Space: Europe Should Shape the Future of Space', Available at: https://www.airbus.com/public-affairs/brussels/our-topics/space/new-space.html

ASD. (undated), 'ASD's Contribution to the Consultation on the Future "EU 2020 Strategy"'.

ASD. (2007), 'ASD Welcomes European Union Decisions Enabling the Galileo Satellite Project to Move Ahead', *Press Release*, 30 November 2007.

ASD. (2016), 'An Ambitious Space Industrial Policy must be the Cornerstone of the "Space strategy for Europe" says European Parliament Sky and Space Intergroup', *Press Release*, 22 June 2016.

ASD. (2020), 'Industry Considerations on Technological Sovereignty: Concept Paper', 15 October 2020.

ASD-Eurospace. (2007), 'Eurospace Facts and Figures, 2007 Edition: The European Space Manufacturing Industry in 2006', 11 May 2007.

ASD-Eurospace. (2014), 'Sharing a Future Satellite Communications Capability for Governmental Purposes', Position Paper.

ASD-Eurospace. (2016), 'Space RDT Priorities 2020: The Incremental Roadmap of Technology Research and Development Activities for Space'.

ASD-Eurospace. (2017), 'Position Paper from Industry on Recommendations Towards More Efficient Public-Private Interactions Regarding Public Procurement of Space Programmes', September 2017.

ASD-Eurospace. (2019), 'Strengthening the European Space Sector Through An Ambitious Industrial Policy: High-Level Guidelines from the European Space Industry'.

ASD-Eurospace. (2020), 'Facts & Figures, 24th edition', July 2020.

ASD-Eurospace. (2021), 'Eurospace Facts & Figures – Key 2020 Facts', July 2021.

ASD-Eurospace. (2022), 'Eurospace – Organisation, Activities & Added Value at Your Service', *Powerpoint Presentation*, February 2022.

Cipriano, Jacques. (2007), 'Intervention by Safran to "Space – a dimension for European Security?"', *Second European Security Round Table*, Brussels, 14 May 2007.

Cox, Robert W. (1996), 'Social Forces, States, and World Orders: Beyond International Relations Theory', in Robert W. Cox (ed.), *Approaches to World Order*, Cambridge: Cambridge University Press, pp. 85–123.

de Concini, Alessandro & Jaroslav Toth. (2019), 'The Future of the European Space Sector: How to Leverage Europe's Technological Leadership and Boost Investments for Space Ventures', *Innovation Finance Advisory & European Investment Advisory Hub*. European Investment Bank. https://data.europa.eu/doi/10.2867/497151

DG DEFIS. (2020), 'Draft Proposal for a European Partnership under Horizon Europe Globally Competitive Space Systems', 27 May 2020.

European Commission. (2013), 'EU Space Industrial Policy: Releasing the Potential for Economic Growth in the Space Sector', COM(2013) 108 final, 28 February 2013.

European Commission. (2021a), 'Action Plan on Synergies Between Civil, Defence and Space Industries', COM(2021) 70 final, 22 February 2021.

European Commission. (2021b), 'Space Technologies for European Non-Dependence and Competitiveness – Guidance Document for Horizon Europe Space Work Programme', Version 2.0, 28 October 2021.

European Commission. (2022), *Commission Contribution to European Defence*, COM(2022) 60 final, 15 February 2022.

European Commission & High Representative of the Union for Foreign Affairs and Security Policy. (2022), *An EU Approach for Space Traffic Management – An EU Contribution Addressing a Global Challenge*, JOIN (2022) 4 final, 15 February 2022.

European Defence Agency. (2018a), 'EU Satellite Communications Market', *Fact Sheet*, 2 July 2018.

European Defence Agency. (2018b), 'Space', *Fact Sheet*, 21 September 2018.

European Defence Agency. (2019), 'EDA GOVSATCOM Demo Project Enters Execution Phase', *Press Release*, 16 January 2019.

European Space Agency. (2005), 'The European Space Sector in a Global Context: ESA's Annual Analysis 2004', May 2005

Galle, Jean-Loic. (2018), 'Letter to Jean-Claude Juncker and Gunther Oettinger', *ASD-EUROSPACE*, 11 April 2018

Gill, Stephen & David Law. (1993), 'Global Hegemony and the Structural Power of Capital', in Stephen Gill (ed.), *Gramsci, Historical Materialism and International Relations*, Cambridge: Cambridge University Press, pp. 93–124.

Hayward, Keith. (2011), 'The Structure and Dynamics of the European Space Industry Base', *ESPI Perspectives* No. 55, December 2011.

Holman, Otto & Kees van der Pijl. (2003), 'Structure and Process in Translational European Business', in Alan W. Cafruny & Magnus Ryner (eds.), *A Ruined Fortress? Neoliberal Hegemony and Transformation in Europe*, Lanham: Rowman & Littlefield Publishers, pp. 71–93.

Morsillo, Giuseppe. (2016), 'Competitiveness Challenges of the European Space Industry', Keynote Speech, Fostering the European Space Economy: Market Conditions for the Space Industry, ASD-Eurospace.

Naujokaityte, Goda. (2021), 'Industrial Partnership on Space Is Grounded – and the Industry Wants to Know Why', *Science/Business*, 30 March 2021. Available at: https://sciencebusiness. net/news/industrial-partnership-space-grounded-and-industry-wants-know-why

Oikonomou, Iraklis. (2013), 'The Political Economy of EU Space Policy Militarization: The Case of the Global Monitoring for Environment and Security', in Anna Stavrianakis & Jan Selby (eds.), *Militarism and International Relations: Political Economy, Security, Theory*, Abingdon: Routledge, pp. 133–146.

Oikonomou, Iraklis. (2017), 'Profits in Orbit: The "Nationalisation" of the Galileo Programme', in Thomas Hörber & Emmanuel Sigalas (eds.), *Theorizing European Space Policy*, Lanham: Lexington Books, pp. 141–158.

Polkowska, Malkorzata. (2020), 'Space Defense in Europe: Policy and Security Aspects', *Polish Political Science Yearbook*, 49 (2), pp. 127–139.

Posaner, Joshua. (2019), 'The Race for Europe's Place in Space', *Politico*, 13 August 2019. Available at: https://www.politico.eu/article/europe-space-race/

Whittle, Mark, Andrew Sikorski, James Eager & Elias Nacer. (2021), 'Space Market: How to Facilitate Access and Create an Open and Competitive Market?', *European Parliament, ITRE Committee*, November 2021.

7 Italy's Space Policy
Between Domestic Preferences and European Policies

Antonio Calcara

Introduction

On December 15, 1964, Italy became the third country – after the Soviet Union's Sputnik in 1957 and the American Explorer in 1958 – to launch a satellite into space. The "San Marco" project symbolises the long Italian tradition and its competitive scientific, technological and industrial base in space. Italy currently possesses the third largest European space industry, in a context where space is now simultaneously a profitable field of economic investment,[1] and also at the centre of international military competition (Oikonomou and Hoerber, 2023). Where does Italy stand in the debate on space from an economic and military point of view? What is Italy's position in the context of European space cooperation?

Based on these research questions, the chapter has two goals: first, it aims to understand what role economic and military motives play in the current Italian space policy, with particular attention to the interactions between the government and the defence and aerospace industry. Second, the chapter investigates Italian preferences in the European context. To achieve these goals, this contribution draws on Liberal Intergovernmentalism (LI) to shed light on the complex interaction between domestic preferences and European policies.

The chapter highlights two fundamental aspects of Italy's space policy and position in Europe: first, Italian preferences have gradually – yet firmly – aligned with those of its defence and aerospace industry. This is due to the concentrated structure of the space market and the powerful incentives of European space cooperation for technological and industrial competitiveness. This is also linked with the changes that have gradually taken place in the EU space policy, where the industrial, economic and defence dimensions are gradually merging into a single policy-area. The key role played by the new European Commission's DG on Defence Industry and Space, as part of DG Industry portfolio, is a first step to solidify a unified leadership in European space policy, with structural links to industrial and defence policies.

Secondly, the Italian position in Europe is characterised by a pragmatic balance between cooperation and competition, between the quest for regional efficiency vis-à-vis global actors and the preservation of national autonomy vis-à-vis European partners. In this context, the United States is viewed in Rome as a convenient

DOI: 10.4324/9781003230670-10

external partner, which can be brought in to correct intra-European imbalances and prevent the prospect of larger states and industries dominating the European space market. As such, Italy is likely to continue sitting on the fence between European efficiency and domestic autonomy, embracing intergovernmental cooperation while resisting fully fledged EU space integration.

This chapter provides two main contributions to the scholarly literature: first, the theoretical framework inspired by LI sheds light simultaneously on domestic preference formation and European space cooperation. Second, this chapter fills an empirical gap in the literature since there are no analyses of the dynamic inter-action between Italian preferences in the European context. In this regard, space represents a unique angle to observe the evolving posture of Italian security policy in times of international competition.

This chapter is structured as follows: the next section presents the theoretical framework and the research hypotheses. The empirical analysis is structured around three main periods: the 1950s–1970s, the 1980s–1990s and the 2000s–2010s. The conclusions discuss the research findings and its theoretical and empirical implications.

LI and European Space Policy

LI aims to explain why and how EU states agree to move integration forward. This approach is based on three main assumptions: first, national governments are the most important actors steering the direction of EU integration. Second, the interactions between political and economic-industrial groups play a central role in determining national preferences (Moravcsik, 1997). Third, relative power among governments and the negotiation tactics of policymakers are key variables to explain integration outcomes. Following these assumptions, the LI approach proposes a two-stage research design. At first, researchers need to identify domestic preference formation that is oriented towards increasing (and possibly maximis-ing) national welfare in the issue area at hand. In the second stage – given different cross-national preferences – the LI focuses on bargaining outcomes as a function of the distribution of interests and capabilities. Governments bargain with their peers to resolve distributional conflicts among competing domestic preferences (Moravcsik, 1998: 18–23).

Recent works that draw on LI suggest that domestic preference formation is a complex political process where different actors compete for political influence (Csehi and Puetter, 2021: 470). Political influence is, in turn, linked with power configurations between political and economic-industrial groups. In the space sec-tor, we can identify two main actors that collaborate and sometimes compete for political influence in a national setting: the government and the industry. On the one hand, national governments need to balance military, industrial and economic considerations; they are aware that space nowadays not only is an integral part of military competition but also has important implications for a country's scien-tific, technological and industrial competitiveness (George, 2019). On the other hand, the space industry is concentrated around a few large players – due to the extremely high financial, technological and knowledge-based market entry barriers – thus creating a substantially oligopolistic market. Space companies ideally aim to maximise their profits and lobby for national space policies that simultaneously protect domestic markets from foreign competition and enable industry's internationalisation and expansion in other markets.

National preferences, then, must find common ground at the European level, characterised by the simultaneous activity of the European Space Agency (ESA) and the EU. ESA, founded in 1975, is responsible for coordinating the collaborative space programmes of 22 European countries. The governance of the Agency is strictly intergovernmental and ESA awards contracts on the basis of the *juste retour* principle, thus allowing each country to receive a financial and technological return for its investments. Within the EU framework – and since the 2004 Framework agreement between ESA and the EU – the European Commission has also assumed important responsibilities in promoting space activities and synergy between civil- and military-oriented space programmes (Oikonomou and Hoerber, 2023). EU institutions directly manage two big cooperative programmes: Galileo and Copernicus. Recently, the European Commission has also created a DG on Defence Industry and Space that specifically implements EU space policy. Unlike ESA, however, EU space activities are an integral part of European industrial policy and the single market and, therefore, not subject to the *juste retour* principle.

The European space policy is driven by simultaneous incentives towards cooperation and competition. Since the 1960s, regional cooperation has been framed as the only way for Europeans to close the "technological gap" with the United States and other global powers. Suzuki listed five of the main reasons underlying European cooperation: (1) to gain economies of scale; (2) to share research and development risks; (3) to increase financial support from partner governments; (4) to increase the market size for the product; and (5) to strengthen European industry in the international competition with much larger American counterparts (Suzuki, 2004: 3). However, the incentives for European cooperation are mitigated by two important factors: first, not every European country possesses a space industry; hence, "have-not" countries may have less interest in cooperating or, in any case, in investing economic and political capital to champion this sector in Europe (Lembke, 2001: 18).[2] Second, European cooperation may lead to distributional conflicts within Europe itself. Different space programmes and industries are also competitors both within and outside the European market and may have the incentive to hamper cooperation if they perceive to be disadvantaged by the terms of cooperation or to protect domestic markets from foreign competition. Balancing cooperative and competitive pressures is key to contextualise member states' preferences towards European policies in sectors that have both economic and security implications (Calcara and Simòn, 2021).

Italy's Domestic Preferences in the European Context

Given these general considerations on domestic preference formation and the European space policy, we can now zoom in on the Italian case. The governmental institutions responsible for space policy are the Ministry of Economic Development for the economic aspects and the Ministry of Defence for the defence-industrial and strategic issues. In this context, the Italian Space Agency (ASI) acts as an intermediary between governmental institutions and industry. Recently, Law No. 7/2018 established the "*Comitato Interministeriale per le politiche relative allo spazio e all'Aerospazio*", an inter-ministerial collegial body that includes all relevant public and institutional stakeholders (Senato, 2018).

The Italian industry has a total turnover of almost €2 billion and more than 200 companies (Ministero dello Sviluppo Economico, 2020). Yet, the market is mainly structured around four leading companies – Thales Alenia Space, Telespazio, Leonardo and Avio – which account for about 80% of the entire space industry. Thales Alenia Space and Telespazio are both subsidiaries of the semi-state-owned Leonardo group and part of a broader Franco-Italian joint venture. Thales Alenia Space is leader in the European upstream space sector, dealing with the construction of satellites and technologies for earth observation, security and exploration of the solar system. Telespazio and Leonardo are specialised in the downstream sector, i.e., the management of satellites and ground services. Avio, a company listed on the stock exchange in 2017, is active in the development of small-medium launchers (the Vega programme). The structure of the Italian space industry is then based on a vast network of SMEs, which has recently organised itself into the National Aerospace Technology Cluster (CTNA).

In order to grasp the Italian preferences in the European context, we rely on the analytical lens of the LI to build two main hypotheses: first, given the importance of the Italian space industry, the concentrated structure of the market and the close relationships between semi-state-owned firms (e.g. Leonardo) and governmental institutions, we hypothesise that industrial interests will be prevalent in shaping domestic preferences. The conjecture here is that the Italian space policy follows similar dynamics of the adjacent defence-industrial sector, where industries have been able to decisively influence the government on the way it approaches interactions with partners and competitors in the global marketplace (Calcara, 2017). We thus expect the space industry to be able to "capture" the government's decision-making process to its own advantages and its proximity to decision-makers offers corporate elites greater chances to influence government preferences.

Regarding Italy's position on European cooperation, several studies have noted how Rome is continuously sitting on the fence between Europeanism and Atlanticism in its foreign and security policy (Cladi and Locatelli, 2021). However, these works struggle to explain when Italy opts for a more Europeanist or Atlanticist stance. Following other works on the political economy of EU defence (Calcara and Simon, 2021), European cooperation can instead be conceptualised as a delicate balance between European efficiency and domestic autonomy. European states are incentivised to cooperate, because they expect to be strengthened vis-à-vis larger global players. At the same time, they also aim to protect their autonomy from their European fellows. In the Italian case, therefore, we hypothesise not only a simultaneous push towards greater European cooperation to strengthen its industrial competitiveness at the global level but also a resistance towards market integration, which could benefit more efficient players with a more integrated industry (e.g. France and Germany). We expect Italy to prefer intergovernmental governance of space, because of the possibility to influence the terms of coopera-tion, to benefit from *juste retour* and to keep possible derogations and exemptions from common rules to avoid possible Franco-German dominance. For the same reasons, we also expect Italy to try to ensure collaborative agreements with the United States and to oppose attempts to full-fledged integration of the space sector at the EU level.

Research Design and Methodology

To test the research hypotheses, this chapter relies on historical process-tracing. This method aims to trace causal process between an independent variable and the outcome of the dependent variable (Beach and Pedersen, 2019). Process-tracing also allows to investigate the alternative causal pathways through which the out-come of interest might have occurred. In this case, the two expectations on domes-tic preference formation and Italy's preferences in Europe will be tested along two alternative hypotheses. First, the argument on industry capture will be set against the prevalence of Italy's strategic-military interests in shaping domestic preference formation.

Second, as regard the Italian position in Europe, our hypothesis will be set against the Atlanticist vs. Europeanist divide in European security. Scholars have argued that the variation between Europeanism and Americanism in Italy's foreign policy can be traced back to the colour of the ruling coalition – with centre-right govern-ments willing to side with the United States, and centre-left governments being pro-European (Cladi and Webber, 2011; Brighi, 2013). We should therefore find that centre-left governments will favour European cooperation, while centre-right governments will prefer to align with the United States. To be sure, the research hypotheses flagged here do not claim to be mutually exclusive or exhaustive and more nuanced explanations may come to light during the analysis (Table 7.1).

The empirical analysis will be structured around three main periods: 1950s–1970s; 1980s–1990s; and 2000s–2010s. At this point, two additional cave-ats are in place: LI has been originally designed to explain the major turning points in the EU integration history, especially treaties' negotiations. In this case, I will focus on the *longue durée*[3] of European space cooperation. This choice nega-tively impacts on the analytical depth of specific moments of space integration,

Table 7.1 Research Hypotheses

	Main Hypothesis	Alternative Hypothesis
Domestic preferences	Industrial interests	Strategic-military interests
Italy's position in Europe	Balancing efficiency and autonomy/cooperation over integration	Europeanism or Atlanticism/ domestic political coalition

while allowing for a broader investigation of possible elements of continuity and change in the medium-long term. Secondly – contrary to conventional LI two-stage design – the analysis will not be presented separately for the domestic and European contexts. Besides space constraints, this is a useful analytical choice to grasp possible interconnections between the two stages. As argued by Csehi and Puetter (2021: 466), the notion of the two-stage design – while analytically useful – becomes increasingly blurred as domestic and EU-level politics are deeply interwoven.

1950s–1970s: From Science to Business

Space entered Italian policy in the late 1950s through the activities of two main actors: the military and scientists.

The first steps were taken by the Italian Air Force through a programme for upper atmosphere research and the creation of a launching range in 1956 at Salto di Quirra, Sardinia (Krige and Russo, 1995: 10). The Air Force Engineer Corps specialised in liquid-propelled rockets and the Italian navy – in collaboration with German scientists – designed a rocket development programme (De Maria and Orlando, 2008: 42). The main leader of the Air Force in space was Colonel Luigi Broglio, who then became one of the main protagonists of Italian space policy.

In the same period, scientists were also showing a growing interest in space research. Edoardo Amaldi, the Italian representative at CERN, aimed to replicate the Geneva organisation and create a European body for space research. Amaldi's idea was to conduct peaceful scientific research, keeping out any military appetites. He underlined that it was "absolutely essential" that the future organisation "has no military connotation and no connection with whatsoever military agency" (quoted in Krige and Russo, 1995: 15). Amaldi then, encouraged by the support of his French counterpart Pierre Auger, addressed a letter titled "Space Research in Europe" to fellow scientists to propose an organisation similar to CERN to be set up for space (CERN, 2012).

European space policy was from the outset characterised by two different and progressively irreconcilable views: on the one hand, a strongly pan-European scientific component; on the other, the awareness that space would also trigger competing military and industrial attitudes. This tension was immediately visible in the 1960s during the development of two European initiatives: the European Space Research Organisation (ESRO) and the European Launcher Development Organisation (ELDO).

On ESRO, Rome clashed with its European counterparts to establish laboratories in Italy (against the view of France and the Benelux countries) and only agreed to join the organisation in exchange for leadership in satellite test vehicles (STV), a project that was very important to Broglio (Caprara, 2012: 177). The debate on the institutionalisation of ELDO was instead characterised by the Anglo-French proposal to co-develop with European partners a joint launcher based on the British military rocket Blue Streak. The project was driven by clear British and French industrial interests and by competitive pressures vis-à-vis more sophisticated

US products (Sheehan, 2021: 102). The Italian reaction to this proposal was particularly cold for two interrelated reasons: first, Blue Streak had already been developed by British military companies and went against the principle that guided organisations as CERN, where researchers and technicians from all countries collaborated from the outset to develop the scientific programme. Secondly, Italian industry would have been disadvantaged by Anglo-French cooperation. In this regard, Amaldi argued that Italian industries would be "excluded from the most important and essential part of the project" (quoted in Mariani, 2015: 195). Italy's non-participation was, however, a problem for the Anglo-French leadership, which aimed to build a multilateral support for their project. The ELDO convention was then signed, after considerable British pressure on Italy, on 30 April 1962, among seven participating states: Australia, Belgium, France, Germany, the Netherlands and the UK. Italy, led by Broglio, entered with a 10% share and several reservations, which would become visible in the following years.

In the 1960s, Italy was consolidating important relations with the United States. On 12 April 1961 – the same day that Gagarin became the first human in space – Broglio presented the San Marco project to the Americans in Florence. This coincidence helped convince the United States – worried about Soviet competition – to support one of their allies in an ambitious space project (Caprara, 2012: 143–144). In 1962, Italy and the United States signed an agreement for the "San Marco" project, and two years later, on 15 December 1964, the first Italian satellite, San Marco-1, was launched by an American Scout rocket. US President Johnson complimented the Italians for the successful launch and for being the first nationality after the Americans or the Soviets to put a satellite into orbit.

In the meantime, European space organisations were revealing some first cracks, which proved to be long-standing obstacles to regional cooperation. On the one hand, Italy constantly complained that industrial return was insufficient and was frustrated that its contribution to ESRO/ELDO went to the benefit of more competitive companies in Britain and France (Bondi, 1993). Between 1968 and 1969, Italy blocked the TD programme and withdrew from Ariane and from ELDO. Italy also opposed the Franco-German telecommunication project Symphonie and launched the first major domestic commercial satellite: the SIRIO project. The newly formed consortium *Compagnia Industriale Italiana* (CIA) won the bid for the procurement. On the other hand, European partners were irritated by close Italo-American cooperation and by Broglio's proposals to use American launchers in Europe (McDougall, 1985). Bondi, the British chairman of ESRO, took on board the demands of medium and small countries by introducing the principle of "juste retour", i.e., fair industrial return, which would become a permanent feature of space cooperation thereafter. Shortly afterwards, the creation of ESA in 1975 marked the prevalence of industrial interests over purely scientific ones, as the new agency was from the outset focused on fostering European industrial projects and industrial return for the main contributors (Suzuki, 2004: 89–93).

In the context of Italian domestic preferences, two themes should be highlighted: first, Italian space policy was characterised by a progressive clash between

Amaldi's idealistic vision and Broglio's pragmatism. While Amaldi was against any involvement of the military or NATO, Broglio emphasised that "military interest in space was so logical and inevitable that there was little realistic prospect of creating an organization without a military dimension" (quoted in Sheehan, 2021: 106). Tensions between the scientific community led by Amaldi and Broglio's San Marco team have probably reduced the possibility of creating a national space agency in the wake of the American NASA or the French CNES.[4]

Second, Italy did not have a consolidated space industry in the 1950s–1970s. The above-mentioned CIA consortium was weakened by the internal rivalry between Aeritalia and Selenia, as both were competing to become the main leader in the Italian space market. This rivalry was, in turn, undermining Italian lobbying to grab European projects. The smaller weight of Italian industries compared to their European competitors, the tension between the scientific and industrial component, combined with European scepticism about Italy's privileged relationship with the United States, led the Italian delegate Carlo Buongiorno to declare during the Blue Streak meeting that he "felt like an ant among elephants" (quoted in Ferrone, 2011: 56–57).

1980s–1990s: Consolidation of the Italian Space Ecosystem

The 1980s and the 1990s brought a decisive consolidation of the Italian space ecosystem and a strong – and from that moment on permanent – alignment between Italian state preferences and the defence and aerospace industry.

In October 1979, the *Comitato Interministeriale per la Programmazione Economica* (CIPE) approved the first national space plan (*Piano Spaziale Nazionale* - PSN) for 1979–1983, which allocated 200 billion *lire* (roughly 100 billion euros) for basic space research and space telecommunications (De Maria, Orlando, and Pigliacelli, 2003: 28). The plan had, as observed by the head of PSN Luciano Guerrero, the dual objective of coordinating a more consolidated industrial base and improving Italy's lobbying within ESA. First, Selenia Spazio was created in 1983, with shareholdings by Selenia, Aeritalia and Italtel (Felice, 2010: 619). In 1990, the constant collaboration in national and European projects led to the merger of Aeritalia and Selenia, and the birth of Alenia Spazio, the first Italian national champion in the field of space industrial production. The company – controlled by the state-owned Finmeccanica (now Leonardo) – brought together the best of the country's existing space capabilities, apart from propulsion (which remained with Fiat-Avio). As noted by De Maria, the creation of Alenia Spazio "resulted in a more balanced Italian participation in the successive European programmes, notwithstanding the fact that Alenia Spazio's size always remained inferior to those of its French and German counterparts" (quoted in De Maria, Orlando, and Pigliacelli, 2003: 27). The PSN also envisaged the launch of two large domestic programmes: the IRIS launcher (later replaced by the Vega), which would make Italy a competitor for medium-small launchers to the larger French (Ariane) and US (Scout) programmes. The second project was ITALSAT, which served as a continuation of the SIRIO programme, and was developed domestically after the rejection in the negotiation phase of the Italian proposal to develop SIRIO-2 with ESA partners.

Secondly, political and industrial representatives were planning the creation of the ASI, with the aim of coordinating a strategic vision of space programmes and guaranteeing greater industrial returns within the ESA. The management of the ASI was characterised by intense inter-institutional rivalry. The Italian space policy was in fact originally structured along two lines, with the Ministry of Research controlling the ESA part and CERN coordinating the national part (including the PSN) (Reibaldi, 1996). As noted by Landoni, however, both the Ministry of Research and the Consiglio Nazionale delle Ricerche (CNR) were deemed to be too centralist and bureaucratic, and a new public agency was needed to ensure not only technical and scientific expertise but also a managerial approach of *liason* with the industry (Landoni, 2015: 35). In 1988, after a long negotiation, ASI was established with the objective of maintaining the competitiveness of the Italian industrial sector. Given ASI's pre-eminence in managing Italian space policy, the Office of Space activities within the Ministry of Research was closed in 1992.

The consolidation of the Italian industrial and institutional ecosystem took place in a period of intense relaunch of European cooperation. The 1986 ESA ministerial conference in The Hague announced the Ariane 5 launcher, the Hermes spaceplane, the Columbus research module and the DRS72 global data collection and transmission satellite. Italy had important industrial stakes in these projects, especially for the development of aeronautical technologies (for Alenia Spazio) and propulsion (for Fiat Avio) in the Ariane and Hermes projects. For the DRS data transmission satellite system, the Italians acted as the prime contractor (Landoni, 2015: 39). The report drawn up by the Italian Research Minister on his return from the Hague conference stressed the need to prevent Italian industry from finding itself in a "subordinate position to France and Germany, thus lacking real competitiveness in strategic sectors".[5] Italian space policy in the 1980s also maintained the privileged relationship with NASA, consolidated through the fruitful collaboration on the San Marco satellite. Luciano Guerrero and Carlo Buongiorno, respectively President and general director of ASI, travelled extensively to the United States to consolidate the bilateral relationship (Ferrone, 2011). In that period, Italy developed in collaboration with NASA the Tethered project, a satellite for research on ionosphere, which was later included in the STS-75 Space Shuttle mission (De Maria, Orlando, and Pigliacelli, 2003: 28).

The consolidation of the Italian ecosystem was starting to bear fruit. The coefficient of geographical return remained below the level of 1 (equality) up to 1988, with a minimum of 0.80 in 1980–1981, but began to climb in the 90s (De Maria, Orlando, and Pigliacelli, 2003: 27). At the 1995 ESA meeting in Toulouse, the Italian and French ministries reached a bilateral agreement to overcome *juste retour* problems in some important projects, including Italy's holding 15% in Ariane, 25% of Columbus and 15% of Hermes (Suzuki, 2004: 127). By the mid-1990s, the Italian space industry had caught up with and surpassed the UK industry in terms of number of employees and turnover and was progressively approaching

German levels (Landoni, 2015: 149). In 1997, ASI developed its first national space strategic plan and aimed to play a more decisive role in Europe. At that time, Italy also engaged in lengthy and successful negotiations with Germany to elect Antonio Rodotà, a Finmeccanica executive, as head of the ESA.

The relaunch of space in Europe through major investments did not, however, eliminate the structural tension between European efficiency and domestic autonomy that characterise European space cooperation. In the late 1990s, ESA, ASI and CNES issued a position paper to reform the governance of ESA, denouncing the inefficiencies of *juste retour* and the veto power of smaller countries (Cheli and Schrogl, 1999). The problems in European space cooperation were also linked to the different European positions' vis-à-vis the United States. The NASA had in fact proposed to the Europeans to participate in the International Space Station, provoking a favourable reaction from Rome and Berlin and a cold reaction from Paris. The French were particularly concerned that the privileged relationship between Italy and the United States could damage European cooperation. As expressed explicitly by ESA Director General Jean Marie Luton, there was "a political problem that Italy has to solve between the United States and Europe" (quoted in Caprara, 2012: 594; author's translation).

2000s–2010s: Opportunities and Challenges of European Space Autonomy

The early 2000s were characterised by a growing military focus on space. The transformation of European armed forces from the protection of national territories to expeditionary out-of-area operations required greater integration of force structures through satellite communications (Borrini, 2006). In 2001, Italy launched its first military telecommunications satellite (SICRAL I), built by the SiTAB consortium: Alenia Spazio (70%) was responsible for the satellite and systems integration, Avio (20%) for propulsion, and Telespazio (10%) for management and control. In the 2000s, the COSMO SKY-MED (Constellation of Small Satellites for the Mediterranean basin Observation) project was the first satellite constellation devoted to the earth's observation for both civil and military purposes (Borrini, 2006). The project consists of four Italian X-band radar satellites and two French optical satellites (PLEAIDES).[6]

The most important project of European cooperation in space is Galileo, a civil satellite navigation and positioning system. The project stemmed from the willingness to find an alternative to the US Global Positioning System (GPS), whose integration into European armed forces had created problems during operations in the Balkans (Salomon, 1999). The project was also strongly supported by European companies, which feared increased competition in this area, since the European market share in satellite navigation in the late 1990s was only around 15% of the European market and 5% of the global market (Lembke, 2003: 262). Italy was a staunch supporter of Galileo, initiated during the Italian presidency of the European Council. Italian policymakers also shared with France the idea that Galileo was strategically important for peacekeeping and peace enforcement operations in the Mediterranean (CEMISS, 2005). However, the Italian position collided with Germany over project leadership. Rome based its claim to the leadership by pointing out that it had never received a leading role in past prestigious ESA programs (Lindstrom and Gasparini, 2003: 17). Germany based its claim on the fact that Berlin contributed the most to ESA. After multiple high-level negotiations between Berlusconi and Schröder, the dispute was finally resolved in May 2003 when it was agreed that Germany, Italy, the UK, and France would each receive a 17.3% and Spain would get 10.3% of Galileo work-share (Lembke, 2003: 260–262).

The Italian industry was, in that period, preparing the ground for further market consolidation. The European aerospace industry had to cope with greater technological complexity and higher costs, a globalised market and competition from US giants. It was therefore imperative for the leadership of Finmeccanica, which had also taken over Telespazio, to look for a partner for Alenia Spazio. There were three alternatives: first, a proposal by the pan-European group EADS (now Airbus Group) to merge Alenia Spazio with Astrium; second, there was the possibility of merging Alenia Spazio with the smaller French group Alcatel; finally, there were bids from American companies, especially from Boeing. The latter option was rejected because it would have created discontent within ESA, as a major European company would have ended up in American hands. The first option was also abandoned for fear of being marginalised with a minority share in the pan-European group (Felice, 2010: 617). In the end, Finmeccanica's management opted for Alcatel, through two different joint ventures: Alcatel Alenia Space, in which Alcatel had 67% and Finmeccanica 33%; and Telespazio with reversed shareholding. In 2007, since Alcatel made over the space portfolio to Thales, the joint venture was renamed Thales Alenia Space. The agreement with Alcatel allowed Finmeccanica to retain veto rights and to maintain an ownership structure with state participation (an option possible with the French, but considered more difficult with Anglo-American groups) (Landoni, 2015: 155).

The period between 2000 and 2020 was also characterised by a gradual EU intervention in space policy. The European Commission issued the "White Paper on Space Policy" in 2003 and the communication on the "European Space Policy" in 2007, which focused on the importance of a competitive space industry and laid the foundations for EU-ESA future agreements. Since 2006, there has been a wide-ranging debate among the European Commission, ESA and the European Defence Agency, culminating in 2011 with the results of a Joint Task Force that identified critical space capabilities that are strategic to the EU's technology supply. In this context, the management of Galileo was a real watershed, given the need for collaborative arrangements between ESA and the European Commission. As noted by Jean-Jacques Dordain, ESA's Director General from 2003 to 2015, harmonising with European standards represented a "headache for the ESA member states accustomed to receive industrial activities proportional to their contributions" (quoted in Lambright, 2016: 509).

European space policy represents both an opportunity and a challenge for Italy. Greater coordination between the state and industry has definitely benefited Italian industry. As noted by former ASI President Roberto Battiston, Italy has registered a

4% positive return from its contribution to the EU space budget and a 1.2% positive return from ESA programmes. Recently, after the ESA ministerial council in Sevilla (November 2019), Italy received projects worth around 800 million euros with two major missions, Rose-L and Cimr (Copernicus Imaging Microwave Radiometer), both with Thales Alenia Space as prime contractor. ESA has also endorsed the project for an Italian small-to-medium size launcher. The VEGA project foresees Italian leadership with Avio (65%), and a participation of other European countries, including France at 12.43% (ESA, 2021).

However, the opportunities are combined with the inevitable challenges of European space cooperation. First, Italy still has some difficulties in coordinating its own domestic preferences in Europe, as witnessed by Italy's double candidacy for the ESA leadership in 2020, which then went to an Austrian candidate, Josef Aschbacher (Pioppi, 2020). Second, there is the ever-present question of Italy's position in the European context dominated by French (and to a lesser extent German) industry. Italy has, for instance, an interest in preserving European attention to the VEGA project, despite acknowledging that European leadership is particularly focused on the French-led ARIANE launcher. The search for greater integration of space industrial policy, around the new DG Defence Industry and Space, could therefore lead to an increasingly concentrated European market and less margin for manoeuvres for Italian industry's niche capabilities.

Conclusions

This chapter employed the LI framework to investigate Italian space policy and its preferences in the European context. The analysis showed that both Italian and European space policies have been characterised by an initial leading role played by scientists which, however, has gradually been replaced by industrial rivalry, both between Europe vis-à-vis global players and within Europe itself. As regards the process of domestic preference formation, the empirical analysis observed a gradual consolidation of the Italian space ecosystem, both from an institutional (through the establishment of ASI) and an industrial (through Thales Alenia Space and Telespazio) point of view. Furthermore, in recent decades, it shows a greater involvement of EU institutions and the co-existence of intergovernmental agreements typical of ESA with EU supranational market and industrial integration.

This chapter highlights two main findings: first, Italian preferences are subject to a strong influence of industrial motives. This is due both to the concentrated nature of a technology-intensive market substantially dependent on public demand and to European cooperation which, since the creation of the ESA, has rewarded industrial consortia with extensive state support. Although the military aspect is clearly present in European cooperation, the empirical analysis does not support the alternative hypothesis that evokes strategic-military interests as decisive in shaping domestic preference formation.

Secondly, the Italian position in Europe is characterised by a pragmatic balance between Europeanism and Atlanticism, between the search for efficiency through European cooperation and the will to preserve national autonomy vis-à-vis

European partners. In this regard, the relations between Rome and Washington have served as a useful counterbalance to periods in which Italy felt disadvantaged vis-à-vis larger and more competitive European space programmes. The link between space and European market integration may render this delicate balance increasingly difficult, as Italy and other medium and small countries are likely to find themselves constrained by common European rules. The empirical analysis showed great elements of continuity in the Italian position in Europe, largely independent from the political colour of the ruling coalition.

The turn towards a more militarised space policy in Europe represents both an opportunity and a challenge for the Italian political, military and industrial establishment. On the one hand, Italy considers space a fundamental strategic field in a rapidly changing regional and international system (Presidency of the Council of Ministers, 2021). Unlike other countries (such as the United States or France), Italy has not yet established a military space command, although close formal arrangements among ASI, the Ministry of Defence and the Army are currently being strengthened (ASI, 2021). From an industrial point of view, Italian big players are also well positioned to take advantage of new space-related military contracts at the national level[7] or cooperative and funding European opportunities.[8] On the other hand, this analysis has clearly shown that national objectives do not always align with those of European partners, in terms of strategic priorities and resulting military and industry-related implications for the space sector.

Italian and European activism in space should be also contextualised in a broader geopolitical scenario. The United States and China have now identified space as a critical domain for economic, technological and military purposes. This has created cascading effects that have led European policymakers to pay greater attention to space. As noted by the ESA director Josef Aschbacher, "Europe has to realize that if we are not investing, we will be left out of this race" (quoted in Posaner, 2021). In addition, the entry of new private actors into the space industry, including the famous examples of Elon Musk's Space X, Jeff Bezos' Blue Origin and Richard Branson's Virgin Galactic, could put further pressure on European investments. European states and institutions are now in a delicate phase of regulating the legal and political framework through which the EU and ESA will co-finance future programmes. Drawing on this work, we can reasonably expect that future European space ambitions will depend on the complex equilibrium between the quest for efficiency to compete with global economic and military actors and the will to preserve some sort of national autonomy in the unbalanced European context.

Notes

1 The global space economy grew by 6.7 % on average per year between 2005 and 2017, almost twice the 3.5 % average yearly growth of the global economy (EIB, 2018). According to recent forecasts, the economic volume of the global space economy could exceed $1 trillion by 2040. See OECD (2020).
2 France, Germany, Italy, UK, Spain and Belgium provide about 90% of European space industry employment (EIB, 2019).

3 I borrowed this term from the French Annales School of historical writing, to refer to long-term historical structures. See Holmes (2003).
4 Centre national d'études spatiales/National Centre for Space Studies.
5 *Relazione Conferenza Ministeriale dell'ESA* (Aja 9–10 November, 1986). Author's translation.
6 Italy-France cooperation has not been problems free. During the 2011 intervention in Libya, Italian armed forces struggled to receive satellite photos from their French counterparts due to different operational priorities. Italy then decided to buy a spy satellite, Optsat-3000, from Israel. See Spagnulo (2020: 218).
7 Thales Alenia Space and Telespazio recently won the contract to build the Sicral 3 secure satellite telecommunications system. See https://www.telespazio.com/en/press-release-detail/-/detail/sicral-3-pr.
8 For instance, the recently approved PESCO project on "Defence of Space Assets". See https://eda.europa.eu/news-and-events/news/2021/11/16/14-new-pesco-projects-launched-in-boost-for-european-defence-cooperation.

Bibliography

ASI. 2009. Intervista a Luciano Guerriero. *ASI*, July 20, Accessed 1/7/ 2021 https://bandiasi.almaviva.it/it/news/intervista-a-luciano-guerriero

ASI. 2021. Firmato Accordo Esecutivo Tra Esercito e Agenzia. *ASI*, September 9, Accessed 23/11/2021. https://www.asi.it/2021/09/firmato-laccordo-esecutivo-tra-esercito-e-agenzia-spaziale-italiana/

Beach, Derek, and Rasmus Brun Pedersen. 2019. *Process-tracing methods: Foundations and guidelines*. Ann Arbor: University of Michigan Press.

Bondi, Hermann. 1993. Crisis and Achievement: ESRO 1967–1971. In Russo, Arturo (ed.) *Science beyond the atmosphere: The history? Of space research in Europe*. Noordwij: ESA-Special, pp. 139–145.

Borrini, Francesco. 2006. *La componente spaziale nella difesa*. Roma: Rubbettino Editore.

Brighi, Elisabetta. 2013. *Foreign policy, domestic politics and international relations: The case of Italy*. New York: Routledge.

Calcara, Antonio. 2017. Italy's Defence Policy in the European Context: The Case of the European Defence Agency. *Contemporary Italian Politics*, 9(3): 277–301.

Calcara, Antonio, and Simon, Luis. 2021. Market Size and the Political Economy of European Defense. *Security Studies*, 30(5): 860–892.

Camera dei Deputati. 2007. *Programma pluriennale di A/R n.SMD 08/2007 sul lancio di un satellite militare denominato SICRAL-1B*. Roma: Camera dei Deputati.

Caprara, Giovanni. 2012. *Storia Italiana dello spazio*. Milano: Bompiani.

CEMISS. 2005. *Galileo vs GPS: Collaborazione o confronto?* Roma: CEMISS.

CERN. 2012. Edoardo Amaldi and the Oringins of ESA. *CERN Courier*, June 2012. Accessed 21/7/2021. https://cds.cern.ch/record/1734810/files/vol52-issue5-p025-e.pdf

Cheli, Simonetta, and Kai-Uwe Schrogl. 1999. Reshaping European Space Activities. *Space Policy*, 15(2): 61–66.

Cladi, Lorenzo, and Locatelli Andrea. 2021. Explaining Italian Foreign Policy Adjustment After Brexit: A Neoclassical Realist Account. *Journal of European Integration*, 43(4): 459–473.

Cladi, Lorenzo, and Mark Webber. 2011. Italian Foreign Policy in the Post-Cold War Period: A Neoclassical Realist Approach. *European Security* 20(2): 205–219.

Csehi, Robert, and Uwe Puetter. 2021. Who Determined What Governments Really Wanted? Preference Formation and the Euro Crisis. *West European Politics,* 44(3): 463–484.

De Maria, Michelangelo, Lucia Orlando, and Filippo Pigliacelli. 2003. *Italy in Space 1946–1988.* Noordwij: ESA-Special.

De Maria, Michelangelo, and Lucia Orlando. 2008. *Italy in space: In search of a strategy.* Noordwij: ESA-Special

EIB. 2018. The Future of the European Space Sector. *European Investment Bank Annual Report.* https://www.eib.org/attachments/thematic/future_of_european_space_sector_summary_en.pdf

EIB. 2019. The Future of the European Space Sector. *European Investment Bank Annual Report.* https://www.eib.org/attachments/thematic/future_of_european_space_sector_en.pdf

ESA. 2021. Vega. *European Space Agency.* https://www.esa.int/Enabling_Support/Space_Transportation/Launch_vehicles/Vega

Felice, Emanuele. 2010. State Ownership and International Competitiveness: The Italian Finmeccanica from Alfa Romeo to Aerospace and Defense (1947–2007). *Enterprise & Society,* 11(3): 594–635

Ferrone, Enrico. 2011. *Carlo buongiorno: Lo spazio di una vita.* Florence: Logisma Editore.

George, Kelly Whealan. 2019. The Economic Impacts of the Commercial Space Industry. *Space Policy,* 47: 181–186.

Holmes, Frederik. 2003. The Longue Durée in the "History of Science". *History and Philosophy of the Life Sciences,* 25(4): 463–470.

Krige, John, and Russo Arturo. 1995. *The story of ESRO and ELDO, 1958–1973.* Tolouse: European Space Agency. Accessed 21/7/2021 https://www.esa.int/esapub/sp/sp1235/sp1235v1web.pdf

Lambright, W. Henry. 2016. Reflections on Leadership: Jean-Jacques Dordain of the European Space Agency. *Public Administration Review,* 76(3): 507–511

Landoni, Matteo. 2015. *Lo sviluppo dell'industria spaziale Italiana: Coevoluzione di imprese e istituzioni nazionali dello spazio, 1969–2007,* Milan: PhD Thesis.

Lembke, Johan. 2001. Galileo: the politics of space and technology meets the politics of security. *Workshop on Global and Regional Economic Security and Integration, The Swedish Institute of International Affairs, Stockholm.*

Lembke, Johan. 2003. Strategies, Politics and High Technology in Europe. *Comparative European Politics,* 1(3): 253–275

Lindstrom, Gustav, and Giovanni Gasparini. 2003. The Galileo Satellite System and its Security Implications, *Occasional Paper* No 44. Paris: European Union Institute for Security Studies.

Mariani, Valentina. 2015. Prime Prove Italiane di Conquista Dello Spazio: Aspetti Tecnici E Politici Internazionali (1950–1961). *Eunomia, Rivista Semestrale di Storia e Politica Internazionali,* 1: 173–202

McDougall, Walter. 1985. Space Age Europe: Gaullism, Euro-Gaullism, and the American Dilemma. *Technology and Culture,* 26(2): 179–203

Ministero dello Sviluppo Economico. 2020. *L'Industria Italiana dello Spazio,* accessed 3/7/2021. https://www.mise.gov.it/images/stories/documenti/Brochure_09112020_versione_web.pdf

Moravcsik, Andrew. 1997. Taking Preferences Seriously: A Liberal Theory of International Politics. *International Organization,* 51(4): 513–553

Moravcsik, Andrew. 1998. *The choice for Europe.* New York: Cornell University Press.

OECD. (2020). *Measuring the economic impact of the space sector,* 7 October. https://www.oecd.org/sti/inno/space-forum/measuring-economic-impact-space-sector.pdf

Oikonomou, Iraklis, and Hoerber Thomas. 2023. *The militarization of space policy – The EU and beyond.* New York: Routledge.

Pioppi, Stefano. 2020. Corsa allo Spazio Europeo. Fuori l'Italia, Ecco Chi Resta Nella Partita per l'Esa. *Formiche.* Accessed 15/07/2021. https://formiche.net/2020/10/corsa-spazio-europeo-italia-candidati/

Posaner, Joshua. 2021. Europe Risks Being 'Left Out' of Space Race with US, China, Says Space Boss. *Politico*, June 3, Accessed 22/07/2021. https://www.politico.eu/article/europe-space-race-united-states-china-european-space-agency-esa-josef-aschbacher/

Presidency of the Council of Ministers. 2021. *Strategia nazionale di sicurezza per lo spazio.* Accessed 23/11/2021. https://presidenza.governo.it/AmministrazioneTrasparente/Organizzazione/ArticolazioneUffici/UfficiDirettaPresidente/UfficiDiretta_CONTE/COMINT/Strategia_spazio_20190718.pdf

Reibaldi, Giuseppe. 1996. Future Italian Space Policy. *Space Policy*, 12(1): 9–11

Salomon, Arnaud. 1999. "A Question of Independence and Sovereignty", *CNES Magazine*, No. 6 (August), pp. 20–21.

Senato. 2018. Relazione Sulle Attività e i Risultati nel Settore Spaziale e Aerospaziale. *Senato della Repubblica*, Roma. Accessed 18/07/2021. https://www.senato.it/service/PDF/PDFServer/DF/347481.pdf

Sheehan, Michael. 2021. *Militarizing outer space*. London: Palgrave Macmillan.

Spagnulo, Marcello. 2020. *Geopolitica dell'esplorazione spaziale: La Sfida di Icaro nel Terzo Millennio*. Roma: Rubbettino.

Suzuki, Kazuto. 2004. *Policy logics and institutions of European space collaboration*. New York: Routledge.

8 The Case of Luxembourg

A New Role for the Melians?

Helen Kavvadia

Thirty years after the end of the Cold War, an observation made in 1945 by journalist Theodore H. White remains as true as it is timely, that the world is always "fluid and about to be remade" (White, 1978, p. 224). At present, numerous important developments are occurring in the EU and the larger international system in the security context, with new threats, capabilities, "battlefields," and players (Kagan, 2008). Against this intricate backdrop, and amid the reshuffling of the world order, which is mainly being driven by incumbent actors, including the United States, Russia, and emerging powerful players, such as China and India, the EU and its member states have been proactive in securing favorable positions. The EU, although an important political and economic "pole," has only recently, since the late 2010s, sought to raise its security profile, in response to calls for increased "strategic autonomy" and "European sovereignty" (Billon-Galland & Thomson, 2018; Lippert et al., 2019; Palacio, 2020). These calls, specifically for strengthening European independence, self-reliance, and resilience, led to the launch of the Conference on the Future of Europe (CoFoE) (EU, 2021) in May 2021, which is expected to last until Spring 2022. The purpose of the conference is to formulate reforms to be introduced to EU policies and institutions in the medium and long term. The inter-institutionally shaped priorities for a strategic agenda include, among other goals, the digital transformation and security of Europe, given that the EU "has citizens to protect, interests to defend, and values and a rules-based international order to promote" (CEU, 2021, p. 1). In the wake of the Afghan crisis in 2021, and following a series of events in recent years that have exposed Europe's vulnerability to external shocks, the EU "must reflect, openly and clear-eyed, on a new stage in collective security and defence capabilities" (CEU, 2021, p. 1).

With security being traditionally determined to a large extent by technological capabilities, and with space playing currently an increasingly important role (Haas, 2015) in two key future priority domains—digitalization and defense—the EU is actively seeking to raise its profile as a space-faring actor, in order to reap dual-use opportunities. In parallel to concerted endeavors being made at the EU level, some of its large member states are also implementing national policies for increasing the dual-use spill-overs of space technology. Some states, such as Germany, are actively pursuing space militarization in the sense of employing space-based capabilities for terrestrial military purposes, while others, such as France, appear

DOI: 10.4324/9781003230670-11

willing to prepare for space weaponization involving the projection of destructive mass or energy forces from, into, or through space by creating the French Space Command in 2019 (Pasco, 2019).

As the vast majority of EU member states are small, it is interesting to examine whether "the strong do what they have the power to do and the weak accept what they have to accept" (Thucydides, 1972, p. 89). Are small EU states content to act as security consumers (Archer, 2016; Schweller, 1994), following larger powers, or do they develop their own space-related ambitions for becoming security contributors (Huntley, 2007)? Academic interest in both space militarization and the role of small states at the global level has been increasing. The nexus of these two research areas remains, however, under-researched and is as such addressed in the present chapter. The chapter focuses on the European context by examining the role of small EU states in the militarization of space, focusing specifically on the Grand Duchy of Luxembourg as a case study. Given the divergent definitions of small states and the lack of a widely accepted set of criteria to demarcate them (Enrikson, 2001), this chapter adopts geographical size as the essential parameter for defining small states (Baker Fox, 1969).

Luxembourg is an interesting case study because, notwithstanding its small size, it has been at the forefront of space development (Nikam, 2019). This is because of the country's efforts to diversify its quasi-mono-sectoral economic model, which is predominantly reliant on the finance sector. Consequently, Luxembourg has strategically "embraced" space as an overarching sector, leading to comprehensive, state-of-the-art technological developments in digitalization and industry 4.0 (LG, 2020a; Rifkin, 2011). In this vein, Luxembourg was the first EU country, and the second worldwide after the United States, to elaborate a *lex specialis* governing the use of outer space resources (LSA, 2017) while working in parallel to systematically create an environment conducive to the cultivation and development of vibrant technological, financial, and academic space-related ecosystems. Additionally, as explained in a following section, the Grand Duchy has been exploiting dual-use space capabilities to expand into space militarization applications, mainly in the reconnaissance, Earth observation, and communications fields. What rational interests have motivated a small country like Luxembourg, which enjoys a "pampered" security environment—namely, being located in central western Europe and being a member of Euro-Atlantic institutions—to make overtures toward space militarization? This question is particularly compelling as Luxembourg's development of defense capabilities seems *prima facie* unattainable or even futile due to the country's inability to achieve security autonomy given its size. Furthermore, Luxembourg does not need such autonomy thanks to its many alliances with larger powers. That said, through space militarization, Luxembourg can compensate for its finite conventional capabilities, which are obviously restricted by its small geographic size. Indeed, space militarization would enable Luxembourg to transcend its size limitations by flexing its economic and technological potential, allowing the nation to effectively punch above its weight (Frentz, 2010; Harmsen & Högenauer, 2021; Huberty, 2011; Nikam, 2019; Stiles, 2018). In this line of argumentation, this chapter focuses on the hypothesis that Luxembourg employs its hard-power space capabilities as a strategic asset to maximize its soft power.

Testing this hypothesis with Luxembourg as a case study can be of interest to academics and practitioners alike, including decision-makers in the political and private sector spheres, by providing insights into the nature, impact, and interplay of different powers and actors. This approach therefore contributes to deciphering the emerging competitive multi-polar system in the militarized space domain. In contrast to the main body of research, which primarily focuses on space militarization by larger actors (Burger & Bordacchini, 2019; Haas, 2015; Huntley, 2007; Papadimitriou et al., 2019; Peter, 2009), this chapter concentrates on small states. This emphasis is important, namely because, in periods of reconfiguration and fluidity at the global level, with disruptive technological advancements occurring in the space security domain, smaller countries must frame their environment and seek to adjust and reposition themselves within the international arena. Furthermore, the study of small states offers comprehensive insights "into the nature of political power in space lacking in the outlooks of larger states, and useful to both the development of an adequate theory of space power and to the satisfaction of individual countries' security needs" (Huntley, 2007, p. 239).

To extend the currently limited scholarly work on space policies followed by small countries, of which small European states have received a mere fraction of research interest (Al Rashedi et al., 2020; Huntley, 2007; Jermalavičius & Lellsaar, 2013; Johnson & Levite, 2003; Paikowsky & Ben Israel, 2009; Sagath et al., 2018; Saperstein, 2021; Tziortzis, 2020), this chapter addresses the research question from a structural realist perspective. Additionally, this chapter responds to calls to better understand "smaller states' outlooks [as they can] be helpful in building a more general space power theory" (Huntley, 2007, pp. 252–253), this chapter addresses the research question from a structural realist perspective. Complementing the literature on the legal and commercialization aspects of Luxembourg's space development, the chapter seeks to reveal the nation's rational interests by analyzing, through a light discourse analysis, primary official documentation of Luxembourg and of European institutions, as well as secondary sources, including scholarly works on space-related topics. In addition, given the absence of scholarly literature on Luxembourg's space militarization endeavors, press articles are employed as a stocktaking method.

This chapter claims that to advance into space militarization, Luxembourg has used the dual-use features of space technology to gain dual-power benefits. While certainly serving its military aims, Luxembourg has relied more on expanding its soft-power gains. Although the strengthening of hard power for the purpose of enhancing soft power appears antithetical, it is refuted by the analytical finding that hard and soft powers are not incompatible at all as they benefit and reinforce each other. Luxembourg has skillfully converted its militarization from a hard-power, material-related, and cost-implying activity to one that is revenue-generating, intangible, and soft-power engendering (Gray, 2011; Hrozensky, 2016; Lambakis, 2001; Pollpeter, 2008). This is because Luxembourg acts as an anchor customer and a moderator for the development of both military and civilian capabilities (Chong & Maass, 2010), building on the political, regulatory, and financial environment of its space sector. The perpetual interactive feedback between the activities of both sides of the space sector, i.e., its dual nature, creates hard power through military prowess and assets, as well as soft power by further-ing economic and technological growth (Nye, 2004; Telò, 2006). As a supplement, through its economic achievements, Luxembourg has strengthened its prestige in techno-nationalistic terms (Carr, 1939; Johnson-Freese, 2007; de Montluc, 2009; Morgenthau, 1960) and influence (Boulding, 1989; Gray, 2011; Kelin, 2006; Long, 2017; Saperstein & Cera, 2021; Sheehan, 2007). In a period of intensive fermenta-tion concerning space-related issues and major changes in the international order, and as a member of the space-farers' "club," Luxembourg has gained an elevated position and enhanced clout.

The rest of this chapter is organized as follows: The next section outlines the theoretical framework after which, in the subsequent section, Luxembourg's space militarization activities and capabilities are described. Next, the research question is addressed, followed, in the last section, by a presentation of the major findings and conclusions.

Scaling Up Soft Power through the Strengthening of Hard Power. An Oxymoron?

The currently observed militarization of space can be ascribed to an emerging multi-polarity, one which is provoking a series of structural changes in the world order. Within the anarchic structure of the international order, multiple players are being confronted with new security dilemmas (Booth & Wheeler, 2007) and the imperative to adjust their capabilities for the purpose of strengthening their offen-sive and defensive power. This transformative process primarily concerns large and middle-sized powers, but not solely. Arguably, smaller countries cannot aspire to self-sufficiency and well-rounded security. In the spirit of self-preservation, they tend to become security consumers beholden to more powerful countries and preoccupied with the formation and consolidation of alliances with these nations (Baker Fox, 1969; Mastanduno, 1998), mainly as shelter-seekers (Bailes et al., 2016; Thorhallsson, 2011). Their "bandwagoning" implies their "rent-paying" contribution to common security in terms of both material and intangible, as well as financial, means. In this sense, a small country, such as Luxembourg, has the obligation to meet international commitments resulting from its participation in the North Atlantic Treaty Organization (NATO) and membership in the EU. In this vein, Luxembourg has developed security initiatives that, instead of enhancing conventional capabilities, privilege space militarization. The central question in this regard is why Luxembourg has opted for space militarization.

To answer this question, structural realism (Waltz, 1979) can be deployed as a useful lens for explaining state and interstate behaviors across different coun-tries of variable sizes. States are typically viewed as "like units," irrespective of their size and domestic regime, within an anarchical international order, within which they function similarly, despite their differentiated capabilities. Particularly concerning space issues, the interests of all states are driven by the "unity of the domain," which creates an equivalence, despite the "stark asymmetries [and capabilities grounded in] the magnified importance of scientific and technological prowess" (Huntley, 2007, p. 255). In these terms, "from a political perspective, it is this narrowing gap between aspiration and what is feasible in space which makes space policies so intriguing" (Hoerber, 2009, p. 412). Additionally, the analytical approach of structural realism purports to explaining national and international behaviors in a "realistic" way, avoiding the subjection of research results to normative standards. This renders the results of case studies replicable and generalizable to other entities, independently of political regime types, levels of interdependencies, and the reliance of multilateral institutions on "hard law." The core insight of structural realism is that states develop their foreign policies in the interest of self-preservation and power, which, collectively, shape the international order, consequently associating structural realism with rational theory (Glaser, 2010). Despite acrimonious disputes over its definition, "power" is widely considered the ability to direct the decisions and actions of others in the interest of one's own interests and intentions (Georgiou, 2008; Nye, 2002; Riker, 1964; Weber, 2012; Wrong, 2017). Although structural realism does not consider states to be rational decision-makers *stricto sensu*, it does claim that states that do not respond rationally to systemic imperatives suffer adverse consequences. Whether rational or reasonable (Brown, 2012), states always act in their own self-interest (Kahler, 1998; Shadunts, 2016), an assertion that can be viewed as the underlying principle of the rational actor model (RAM), one of Allison's three frameworks of foreign policy analysis (Allison & Zelikow, 1999). Whether through domination or moderate activism, both of which are concepts integral to structural realism, both offensive (Brown et al., 2004; Elman, 2004; Gilpin, 1981; Mearsheimer, 1990) and defensive (Grieco, 1990; Jervis, 1978; Van Evera, 1998; Walt, 1987; Waltz, 1979) realism regard power and interest as the most important determinants of state behavior, shaping, among other phenomena, security issues. However, "one distinction between these two versions of realism is … whether [the anarchic ordering system] encourages states to maximize their security or to maximize their power and influence" (Lobell, 2017, p. 1). Grounded in one of the generic definitions of power, as "the production, in and through social relations, of effects that shape the capacities of actors to determine their circumstances and fate" (Barnett & Duvall, 2005, p. 39), the notion of influence is incorporated into the concept of power. In this sense, it broadens the ways in which power has been traditionally understood beyond its association with a hard, coercive, and material nature, principally emphasizing military power (Mearsheimer, 1990; Waltz, 1979).

Consequently, the concept of power has evolved to encompass a social, cooperative, and intangible dimension (Boulding, 1989; Cohen, 2008; Finnemore & Goldstein, 2013), which Joseph Nye crystallized in 1990 into what later became known as "soft power" (Nye, 1990a). Initially, Nye repudiated realists' views (Nye, 1990b), despite having drawn upon their understanding of power (Digester, 1992). Nye rejected realism notwithstanding the acknowledgment by Kenneth Waltz, the originator of structuralist realist theory, that states actors should not rely solely on material power (Waltz, 1979) and despite the neorealist concept of latent power, which recognizes the socio-economic ingredients of hard power (Mearsheimer, 2016). For several scholars, therefore, "the traditional distinction between hard power and soft power is not entirely persuasive" (Treverton & Jones, 2005, p. xi). While remaining loyal to the cooptive principle of soft power, as a means of "getting others to want what you want... [which is associated] with intangible power resources such as culture, ideology, and institutions" (Nye, 1990b, p. 167), Nye ultimately recognized that soft power is one of the three sources of power—military, economic, and soft—that actors employ to different degrees, and in different proportions, to accomplish their goals (Nye, 2004), ultimately culminating in Nye's self-characterization as a "liberal realist" (Dario, 2020). Nye's notion of soft power was further concretized under mounting criticism concerning the lack of a theoretical framework (Bakalov, 2019; De Martino, 2020; Lee, 2009) and the absence of adequate analytical tools and operationalization mechanisms. Thereafter, the concept evolved beyond "the attractiveness of a country's culture, political ideals, and policies" (Nye, 2004, p. x) to progressively encompass additional dimensions, such as influence and the projection of national power (Harris, 2018), standing, reputation, prestige, and national pride (Krige, 2007; Lee, 2011; Luzin, 2013), influence (De Martino, 2020; Harris, 2018; Lee, 2011; McClory, 2010), leadership (Lee, 2009; Nye, 2017), and information and technology (Lee, 2009; Nye, 2014). Whereas economic strength is not necessarily paired with soft power, as it can be converted into hard or soft power (Nye, 2006), it can be seen as a dimension of soft power, inasmuch as it relies on economic resources, while also being instrumental for economic gains (Carminati, 2021), as actors can "woo [other actors] with wealth" (Nye, 2006). Although "the critical distinguishing factor [of soft power and hard power is] coercion versus attraction ... it is appropriate to regard the two kinds of power as mutual enablers" (Gray, 2011, p. ix) and as complementary (Al-Rodhan, 2019; Nye, 2004) types of power involving "broad national interests that include political, strategic and economic dimensions" (Melissen et al., 2011, p. 4). This is because soft power "can be as important as the exercise of hard power to achieve a nation's desired objectives" (Lee, 2011, p. 11).

With the potential for the dual use of military and civilian applications, space exemplifies the complementarity of the two types of power and, thereby, constitutes an arena at their nexus (Huntington, 1996; Luzin, 2013; Peter, 2009). Employed by "different actors at different degrees [and] in different relationships" (Nye, 2004, p. 30), the two types of power leave unique and individual footprints in space policies. In this manner, states can be viewed as occupying any place along a continuum from pure soft power at one end to pure hard power on the other (Bakalov, 2019; Fan, 2008; Gallarotti, 2011; Rothman, 2011). Ultimately, where a state is positioned on this continuum, as well as the footprint it leaves with its space policy, depends to a large extent on its size, with smaller states tending to opt for soft power methods to achieve their general strategic imperatives (Gray, 2011; Lee, 2011; Treverton & Jones, 2005), as well as those more concretely related to space issues. Approaching the militarization of space through a lens that includes considerations of both hard and soft powers extends the scope of research in this domain to embrace the totality of the concept of power, while concentrating the research focus on a small country broadens the analysis to incorporate both small and middle powers, which constitute the majority of nations not just in the EU but worldwide (Bilgin & Elis, 2008).

Toward a Space Odyssey or a Space Iliad?

Luxembourg joined the space-faring nations in 1985 through the establishment of the Société Européenne des Satellites (SES), a satellite and terrestrial telecommunications network provider, during the third space development cycle, characterized globally by, among other elements, an increased focus on commercial telecommunications (OECD, 2019). Luxembourg's space endeavors have been characterized as an "odyssey" (PWC, 2017). Through brief and targeted stocktaking, and via an analysis of space development in Luxembourg, this section seeks to determine whether the country's space endeavors can be best characterized as an Odyssey or an Iliad, in the sense that, while the Odyssey is the story of a journey and a discovery, the Iliad is an epic tale of achievements on the battlefield. The first concerns the civial use, while the second the militarization of space.

Established by the initiative and support of the Luxembourg government (Higgins, 2007, p. 468), the SES has developed into a major telecommunications satellite network operator, with the state remaining a primary shareholder, making a direct capital injection of 11.6% and a supplementary injection of 21.73% through two public lenders, giving the state a total share of 33.33%. In 2016, in an effort to replicate the success of the SES, Luxembourg launched a fresh impetus, NewSpace, through the SpaceResources.lu initiative, positioning the country as the European hub of space resource utilization (Brennan, 2019; Machuron, 2022). To regulate this new space-related realm, Luxembourg adopted a legal framework for the exploration and use of space resources, safeguarding the rights of private operators and thereby gaining international attention (LSA, 2017). Through concerted efforts, Luxembourg promoted the development of the space sector as a "strategic decision" (Machuron, 2022). Consequently, Luxembourg's industrial policy is aimed at leveraging the space sector, both upstream and downstream, and encompasses research, education, the production of assets and services, as well as finance, "using the same model which caused the financial sector boom" (Sheetz, 2017, p. 1). For this purpose, the Grand Duchy launched the Luxembourg Space Agency (LSA) in 2018 as a coordination hub, which, contrary to other national space agencies, does not conduct research or launch space missions. Rather, the LSA manages national space programs and acts as a facilitator with the mission to accelerate the emerging innovation-oriented businesses (LSA, 2019). Via the country's space ecosystem, the LSA promotes the vertically integrated production of space assets and the horizontal diversification of space activities by public and private stakeholders (LG, 2018). As with the SES, Luxembourg continues to collaborate directly with companies of particular importance, as in the more recent case of Euro-Composites S.A., in which it participates directly and indirectly to produce composite materials for the aerospace, satellite, rail, and maritime sectors (LG, 2019b). Furthermore, in order to attract foreign investment, talent, and clout, Luxembourg "showcases" its space ecosystem at an international level through economic missions, participation in major international events, such as the International Astronautical Congress, and the organization of yearly events, including "New Space Europe" and the "Space Forum." As a result of such concerted space promotion efforts and due to "strong public support" (Sheetz, 2017, p. 1), in 2019, the space industry represented 2% of Luxembourg's GDP (Trouillez, 2019), with about 70 companies and research bodies listed in the space directory in 2021 (Machuron, 2022).

To increase the space industry's GDP share to 5% in the next ten years (Sheetz, 2017), Luxembourg has implemented a space industrial policy (LSA, 2019) based on four pillars: (i) expertise development through the state's anchor customer activity, for spurring sectoral growth through technological development and the expansion of national players beyond its borders via a specialized program (LuxIMPULSE) implemented in partnership with the European Space Agency (ESA), of which Luxembourg has been a member since 2005; (ii) innovation promotion through the establishment of the European Space Resources Innovation Centre as a center of expertise in space-related scientific, technical, business, and economic domains ultimately aimed at human and robotic exploration and utilization of space resources (LG, 2021a); (iii) the acquisition and cultivation of skills and expertise through postgraduate interdisciplinary education at the University of Luxembourg (LG, 2019a); and (iv) funding aimed at bolstering start-ups and early-stage companies through the venture capital investment fund Orbital Ventures, in which Luxembourg has invested "an undisclosed amount" (Foust, 2020) in its capital of USD 140 million (LG, 2021b).

Internationally, Luxembourg projects the civil dimensions of its space capabilities through advertorials that present the country as a "trailblazer on the international space [...] thrilled to join forces with entrepreneurs, businesses, agencies and investors to bring forward the NewSpace industry in general and the space resources utilization sector in particular" (Luxinnovation, 2021, p. 1). Nonetheless, in parallel to this projection, the military use of its space sector has been scaling up, placing Luxembourg among the 12 EU states engaged in space militarization activities and defensive space programs (de Montluc, 2012). In line with the international trend of public–private partnerships in some critical areas of space applications (Anderson, 2020), Luxembourg's militarization of its space sector is occurring both directly and indirectly through official and unofficial channels as well as upstream and downstream developments at the national, European, and international levels within its overall space sector framework, as depicted in Figure 8.1.

Directly, and as part of its defense strategy, Luxembourg promotes:

Developing expertise and capabilities in the future oriented areas of 'space' and 'cyber defense' in order to meet the need for surveillance, communications and data link capabilities, but also the need for data storage and analysis capabilities, [in order] to enhance the safety of [...] military personnel, particularly on deployments.

(LG, 2017, p. 7)

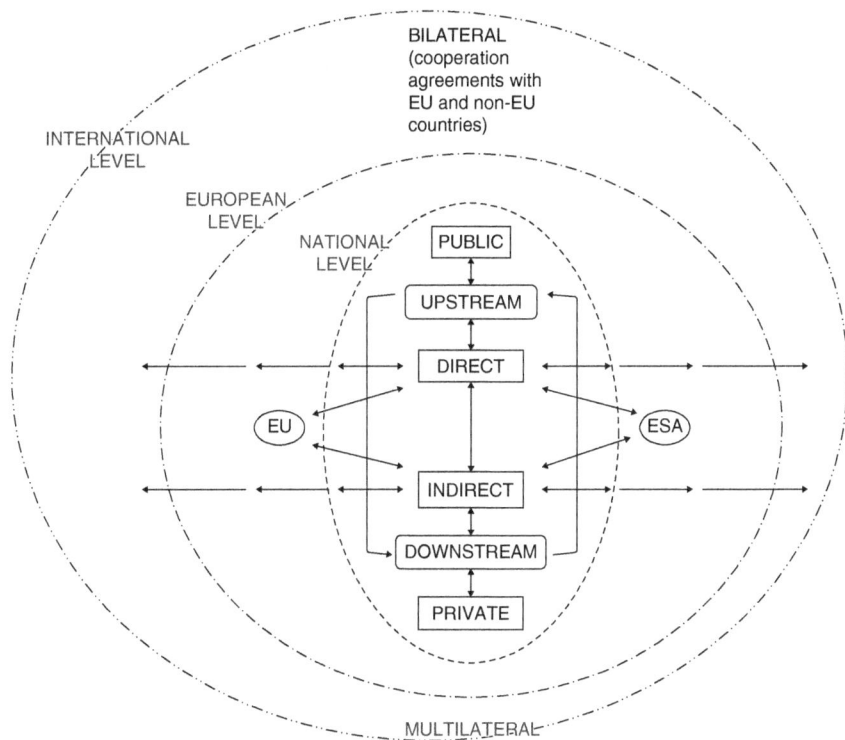

Figure 8.1 Luxembourg's Space Sector Framework.
Source: Author's elaboration.

For this purpose, in 2018, Luxembourg established, with SES, GovSat, a joint venture providing end-to-end encryption of satellite communication (satcom) services. With applications in areas, such as connectivity for theaters of operation, border control, and intelligence, surveillance, and reconnaissance (ISR) (GovSat, 2022), GovSat addresses defense customers, including governments, such as those of Belgium and the United States (Oglesby, 2021b), as well as institutional users, such as NATO, for its forces in Afghanistan (Nikam, 2019), and the United Nations (UN), for its stabilization mission in Mali (Bauldry, 2020). Additionally, also in 2018, Luxembourg began developing public space military assets, such as the Earth observation satellite LUXEOSys, which has been viewed with "great interest" (Schnuer, 2021b, p. 1) by Luxembourg's allies, including the EU, the UN, and NATO. The LUXEOSys program was created by law in 2018, authorizing the government to acquire, launch, and operate a satellite and its ground segment for the stated purpose of Earth observation. However, the project was hampered by several shortcomings in its design, as the army "did not have the infrastructure" (Oglesby, 2021a, p. 1), or because of "miscommunication" (Schnuer, 2020a, p. 1), or even due to "blantant amateurism" (Schnuer, 2020c, p. 1). As a result, LUXEOSys experienced cost overruns of 82%, from its initial budget of EUR 170 million to nearly EUR 309 million as re-estimated in 2020 (Schnuer, 2020a). Nonetheless, interest in the system has been expressed by the US Space Force (Schnuer, 2021a). While the cost overruns of the project are under investigation by the Court of Auditors for, among other potential fraud, Luxembourg has partnered with LUX-EOps, a formal Luxembourg Temporary Partnership founded by Luxembourg-, Belgium-, and Germany-based enterprises (LG, 2020c). LUXEOps will assume, on behalf of and under the direction of the government, end-to-end responsibility for various phases of the LUXEOSys program, from pre-launch preparation and validation of the ground segment to 24/7 in-orbit operations of the satellite platform and payload to the management and "exploitation" of imagery.

Indirectly, Luxembourg has ramped up its space militarization first through its *private sector* activities. Starting in 2005, the SES, more than a third of which is state-owned, has led the way as a provider of connectivity solutions and services "to governments all over the globe, and in particular, the US Government and military" (SES, 2022, p. 1). The government segment, concerning civilian and defense-related applications serving 30 countries, 15 agencies, and more than 60 organizations, including NATO and the UN, underwent considerable growth. As one of the world's leading fixed-satellite service operators (Venet, 2011) of a growing constellation of medium Earth orbit (MEO)-geosynchronous equatorial orbit satellites for ISR, the SES's government sector constituted 38% of the group revenues of the SES in 2020 (SES, 2021a, p. 22), up from 12% in 2017 (SES 2018a, p. 22). The United States, which is responsible for 60% of the activities of the SES's government sector, is served mainly through an SES-wholly owned US-based branch, the Government Solutions (SES GS), which "supports the US Army in conducting a series of cutting-edge trials and testing of commercial satellite constellations [mainly because] MEO satellites are unique in their capabilities and SES operates the world's only commercial MEO satellite constellation" (Ombredane, 2021, p. 1). The SES's cooperation with the United States for military operations has been strengthened further, both upstream and downstream, through a number of initiatives. Examples of these initiatives include contractual cooperation with the US Air Force Research Laboratory to produce terminals capable of military access to the O3b constellation, a satellite-based communications system (Gerhardstein, 2020), and the Hydra platform, which permits the provision of customizable situational awareness information in real time (SES, 2021b). Another Luxembourg company, established in 2017, Kleos Space S.A. (Kleos), provides Luxembourg with "meaningful capacities for [its] defence but also society" (Schnuer, 2020b, p.1) as an artificial intelligence (AI)-enabled radio frequency reconnaissance provider of data as a service for global intelligence and geolocation. With three of the four planned nanosatellite clusters scheduled for launch by the middle of 2022, Kleos will monitor key areas, such as the Strait of Hormuz, the South China Sea, and the arctic and Antarctic regions. With plans to establish a constellation of 20 such clusters in the long term (Schnuer, 2020b), Kleos will provide services to the Japanese military for the identification of illicit activities in territorial and international waters (AA, 2021), which could evolve into "offensive concepts for an AI-enabled battlespace" (Layton, 2021, p. 1). Kleos cooperates with another Luxembourg-based company, Spire Global Inc. (Spire), which operates in the same fashion. With an AI-enabled constellation of roughly 90 nanosatellites and space-to-cloud analytics, covering approximately the same areas as those monitored by Kleos, Spire is cooperating with the Australian Government's Office of National Intelligence (Spire, 2022) and Japan to implement a space-based Automatic Dependent Surveillance–Broadcast surveillance system to generate data specifically tailored for improving the global tracking of aircraft (Spire, 2018).

Second, being committed to and actively supporting multilateralism (Harmsen & Högenauer, 2021) and Euro-Atlanticism (LG, 2017), Luxembourg is carrying out its space militarization activities within EU, European, and multilateral *institutional activity* frameworks. Founded "on broader structures" (LG, 2017, p. 24), Luxemburg's defense apparatus participates in the acquisition and development of programs for common capabilities, which enable contributions to the priority requirements of the EU, the ESA, and NATO. The institutional space militarization activities of the Grand Duchy are channelled through public and/or private sector entities, mostly in a combined manner. Despite not developing security programs, unlike other space agencies (Kolckynski, 2018; Papadimitriou et al., 2019), the LSA coordinates a wide range of institutional activities for meeting national and allied "critical strategic requirements in the field of communications and observation" (LG, 2017, p. 38). Since its launch in 2018, the LSA represents Luxembourg in ESA and EU space-related programs, whilst in parallel, Luxembourg "participates in the Alliance Ground Surveillance (AGS) programme" (LG, 2017, p. 38) and actively explores "opportunities in the space sector, from a European and transatlantic perspective, particularly within the Wideband Global satellite communications System (WGS) programme" (LG, 2017, p. 42). In the future, the LSA plans to acquire remotely piloted aerial system capabilities in the service of its ISR mission in order to further strengthen its "space technologies for communications and observation" (LG, 2017, p. 37) and accommodate "a NATO programme to protect satellites against harmful debris and spot hostile activity in space" (Lambert, 2021b, p. 1). In this vein, Luxembourg participates in ESA programs covering, among other tasks, navigation and Earth observation and EU dual-use space-related projects, including global navigation satellite systems (GNSS), such as Galileo, European Space Surveillance and Tracking, Govsatcom, and Copernicus earth observation. For managing and operating Galileo's GNSS, the SES delivers an array of services (SES, 2018b). The examples above demonstrate Luxembourg's active support of the EU's Global Strategy for security and its contributions "to the measures and structures designed to exploit the full potential of the Treaty of Lisbon as regards the Common Security and Defence Policy (CSDP), particularly in the context of Permanent Structured Cooperation (PESCO)" (LG, 2017, p. 30).

Third, as part of its defense policy, Luxembourg engages *bilaterally* with its allies, which, beyond the United States and Belgium are primarily those countries labeled "key partners," namely, neighboring France, Germany, and the Netherlands, with which Luxembourg deploys its military capabilities "almost exclusively" (LG, 2017, p. 16), incorporating them into joint multinational groups, of which

GovSat constitutes a prime example of the "collective defence capabilities of [Luxembourg's] allies" (Pultarova, 2017, p. 1). Furthermore, from a global perspective, Luxembourg already provides, and plans to increase its provision of, dual-use space capabilities in the framework of its official development assistance (ODA) (Machuron, 2022), as in the case of Burkina Faso. For improved connectivity, Burkina Faso has partnered with the SES to integrate its existing terrestrial networks of communications with the dual-use O3b satellite system (SES, 2019). Such partnerships and commercial approaches typify Luxembourg's space development as a mixture of Odyssey-like discoveries and Iliad-like "epic" strategies.

Evolving from Melians to Athenians?

In the contemporary booming space sector, Luxembourg is seeking to emerge as a European focal point, one "thrilled to join forces with entrepreneurs, businesses, agencies and investors to bring forward the NewSpace industry" (Luxinnovation, 2021, p. 1). Departing from—but alongside—the economic domain, which normally occupies the forefront of the national stage, Luxembourg is strengthening its space militarization activities, contrary to the usual process, whereby civilian applications stem from the military use of space capabilities (Kolovos & Pilaftsis, 2015). Why is a small European state, nestled in multiple Euro-Atlantic security structures, turning to space militarization? Is it an effort to escape the fate of small powers, as the more powerful Athenians reminded the weaker Melians during the Peloponnesian War, more than two millennia ago?

As the context shapes the space priorities of countries (Hoerber, 2009), when examining the question from a structural-realist perspective, geopolitical developments, with new actors and new threats, have to be considered. In this manner, the role of the state as a rational actor pursuing national interests and power in world politics has been emphasized. To explain Luxembourg's space militarization activities, this chapter relied on the RAM, Allison's trademark analytical tool, for "recounting the aims and calculations" (Allison & Zelikow, 1999, p. 13) of these activities. The RAM was applied via a contextual approach to decipher, through a light discourse analysis of official documents and government public statements, the implications of Luxembourg's space militarization activities with reference to the four core concepts of the RAM: goals and objectives; alternatives; consequences; and choices.

The security *context* has changed at the global and European levels since the 2009 Treaty of Lisbon, which hastened the EU's development as a security-policy actor and imposed trans-Atlantic demands on the EU to assume "more of the burden for its own defense" (Kolczynski, 2018, p. 214). The new security context and the allocation of significant EU funding for military space programs (Kolczynski, 2018) constituted a window of opportunity for Luxembourg to accomplish its *objective* of expanding its nascent space industry into the military realm. This expansion was a strategic *choice* facilitated by the fact that Étienne Schneider, the "architect of Luxembourg's space sector" (SWG, 2020, p. 1), was the Minister of Defense and the Minister of Economy when "Luxembourg was looking for ways to diversify its economy and explore new sectoral avenues" (Carey, 2017, p. 2). An integrated policy as a *consequence* could serve both security and the economy (Robinson & Mazzucato, 2019) through the cross-development of both sides of dual-use space technologies for the purpose of reaping dual benefits. Alternatively, and in order to meet its 2014 commitment to NATO to increase its defense efforts and achieve the military spending target of 2% of the GDP by 2020 (LG, 2017), Luxembourg could strengthen its conventional military capabilities. Nonetheless, the chosen synchronous approach to developing space militarization and the economy has allowed Luxembourg to overcome the technological and financial barriers (Wiberg, 1987), often faced by other small states when pursuing access to space by blending the private sector, bottom up, with the security sector, top down, jointly developing revenue-generating civilian uses and cost-implicating military uses. Although in this manner, Luxembourg increased its defense spending from 0.4% in 2019 to almost 0.6% in 2021, the country remains "a defense laggard" (Dalesio, 2021, p. 1), with just one-third of the average defense budget in the EU. Even though the Grand Duchy of Luxembourg plans to increase investment in space capabilities to 0.72% by 2024 (Lambert, 2021a), the country will still remain well behind in its commitment.

Space militarization has enabled Luxembourg to strengthen three of its prime interests:

i *security* at the national and alliance levels, in line with its defense policy, which advocates "capabilities that are relevant in the broadest possible range of situations" (LG, 2017, p. 38). Space reconnaissance and observation can be built on their conventional forms, in which Luxembourg has traditionally specialized, and thereby allow the country "to outgrow its land dimension" (LG, 2017, p. 37). Furthermore, by primarily embracing "the field of satellite communications" (LG, 2017, p. 42), space-related ISR capabilities "shadow technological evolutions that fall within priority areas for Luxembourg, [...] in the domains of military action of intense developments" (LG, 2017, p. 37);

ii *economic growth* for defending both "economic prosperity and security interests" (LG, 2017, p. 12) through "a strategy for industry, innovation and research in order to involve Luxembourg's economic fabric in defence capability building" (LG, 2017, p. 7). Additionally, the country plans its capability increase and diversification to "be implemented in a resolute and intelligent manner [in order to] draw on expertise from EU countries" (LG, 2017, p. 25), creating significant growth spill-overs and embedding security considerations in socio-economic objectives (Karampekios & Oikonomou, 2018);

iii *soft power* in the form of image through techno-nationalistic prestige as "space power has consequences (and profound implications) both domestically and internationally and gives additional overall national power to a State" (Peter, 2009, p. 3). In this sense, Luxembourg's defense policy goal beyond the 2025 horizon is to make a "visible" (LG, 2017, p. 3) defense effort in a way that "the evolution of the armed forces [...] reflect the image of [the] country–integrated with the international community, in full command of modern technology, receptive to research and development and open to EU citizens" (LG, 2017, p. 37). Despite addressing distinct needs, all three interests served by Luxembourg's space militarization converge on the goal of increasing the country's soft power: On the one hand, through its security interests, Luxembourg aims to meet its commitments as a credible and "reliable" (LG, 2017, p. 49) partner, thereby increasing its voice and influence within the various Euro-Atlantic security structures, rather than being security autonomous; on the other hand, economic growth spurs Luxembourg's influence, as space capabilities and innovative technologies allow the country to participate in setting agendas and standards as a member of the "club" of space-faring nations. Additionally, space-led growth can increase the country's outreach to other countries through cooperation agreements, ODA, and paradigm setting, especially among small states.

Luxembourg's real interests, served through the military use of space, are no different than those of the United States (Hayden, 2005), albeit having different footprints on the hard–soft power continuum. Certainly, therefore, Luxembourg's capability strengthening and soft power increase cannot transform a small country into a big player (Haass, 2008). Melians cannot become Athenians through an upgrade to power or status (Saperstein & Cera, 2021). Yet, viewing power as a resource (Baldwin, 2002) or in relational terms (Long, 2017; Rothstein, 1968; Wivel, 2010), Luxembourg's overall power increases alongside its capacity to achieve its intended results, especially by "leveraging the 'soft-power' dividends of space programs in foreign policy" (Luzin, 2013, p. 26). As technology trends increasingly enable access to space capabilities, smaller states become increasingly dependent on the strategic use of the space domain. Nonetheless in parallel, new vulnerabilities, dependencies, and risks emerge, as space capabilities can expand opportunities for collaboration but they can also provoke competition and potential conflict (Johnson & Levite, 2003).

Conclusions

Important developments in the international system and in the EU include radical changes in the security context, with new threats, capabilities, "battlefields," and players. In the new, fragmented geopolitical structure, power dynamics and new technologies create windows of opportunity for different actors, especially small states, such as Luxembourg. Against this backdrop, Luxembourg, which is at the forefront of space development despite its geographic size, has moved into space militarization as a corollary to the dual use of its space capabilities. By strengthening and diversifying its security policy with the addition of a militarization dimension to its space capabilities, the Grand Duchy of Luxembourg has strengthened three prime interests—namely, security, economic growth, and soft power. Increased security through space capabilities not only allows the country to meet its international commitments vis-à-vis the EU and NATO but also helps it to consolidate its image as a "reliable partner." The economic growth stimulated through space militarization diffuses significant growth spill-overs serving both economic prosperity and security interests. Whereas economic might is not necessarily paired with soft power, it can be seen as one of its dimensions, inasmuch as economic resources bolster the projection of soft power while generating further economic gains in turn, facilitating and reinforcing the country's outreach efforts. Soft power, in the form of visibility, prestige, standing, reputation, and national pride, has increased Luxembourg's influence at the international level. Despite addressing distinct needs, all three interests served by Luxembourg's space militarization converge on its goal of increasing its soft power. In this sense, by increasing its hard power, Luxembourg has predominantly gained soft power, while in parallel stimulating economic growth.

Building on its space-enabling environment—including a sophisticated regulatory framework in this area as well as vibrant technological, financial, and academic ecosystems—Luxembourg relies to a significant extent on the private sector for developing its space militarization capabilities, thereby acting as an anchor customer. Having strategically selected space militarization as the way to meet its increased international obligations in the future, Luxembourg meets this objective with limited public spending, having, on the contrary, increased revenues as a result of the space-generated economic growth. Additionally, by joining the "club" of space-faring countries, Luxembourg can make use of its small geographic size to perform a credible partner role, further strengthening its position and clout in the European and international contexts during the contemporary era of major changes in the geopolitical order and the intensive fermentation of space-related issues. Yet, Luxembourg's augmented militarized space capabilities and amplified soft power are not sufficient to upgrade its power in a way that would transform it from a small to a big player, nor can it—at least as of yet—dispense with "bandwagoning" and shelter-seeking. That said, "Any particular small state may not be an 'essential actor' in the system, but [the world] would not be the same [...] without this class of powers" (Baker Fox, 1969, p. 752).

Bibliography

Allison, G., & Zelikow, P. (1999). *Essence of decision: Explaining the Cuban missile crisis* (2nd ed.). Pearson.
Al-Rodhan, N. (2019, May 23). A neuro-philosophy of global order: The case for symbiotic realism, multi-sum security and just power. *APA Online, APA Blog.* https://blog.apaonline.org/2019/05/23/a-neurophilosophy-of-international-relations-the-case-for-symbiotic-realism-multi-sum-security-and-just-power/.
Al-Rashedi, N., Al Shamsi, F., & Al Hosani, H. (2020). UAE approach to space and security. In K. U. Schrogl (Ed.), *Handbook of space security* (pp. 629–660). Springer.
Anderson, G. (2020, July 23). The private sector's role in space exploration. *Real Clear Markets.* https://www.realclearmarkets.com/articles/2020/07/23/the_private_sectors_crucial_role_in_space_exploration_499793.html.
Archer, C. (2016). Small states and the European security and defence policy. In R. Steinmetz & A. Wivel (Eds.), *Small states in Europe. Challenges and opportunities* (pp. 47–62). Routledge.

Asian Aviation (AA). (2021, July 29). *Kleos partners with Japan Space Imaging Corp.* https://asianaviation.com/kleos-partners-with-japan-space-imaging-corp/.

Bailes, A. J. K., Thayer, B. A., & Thorhallsson, B. (2016). Alliance theory and alliance 'shelter': The complexities of small state alliance behaviour. *Third World Thematics: A TWQ Journal, 1*(1), 9–26.

Bakalov, I. (2019). Whither soft power: Divisions, milestones, and prospects of a research programme in the making. *Journal of Political Power, 12*(1), 129–151. https://doi.org/10.1080/2158379X.2019.1573613.

Baker Fox, A. (1969). The small states in the international system, 1919–1969. *International Journal, 24*(4), 751–764.

Baldwin, D. A. (2002). Power and international relations. In W. Carlsnaes, T. Risse & B. A. Simmons (Eds.), *Handbook of international relations* (pp. 177–191). Sage.

Barnett, M., & Duvall, R. (2005). Power in international politics. *International Organization, 59*(1), 39–75.

Bauldry, J. (2020, January 14). *Stabilising Mali: Lux provides satellite and training capabilities.* Delano. https://delano.lu/article/delano_stabilising-mali-lux-provides-satellite-and-training-capabilities.

Bilgin, P., & Elis, B. (2008). Hard power, soft power: Toward a more realistic power analysis. *Insight Turkey, 10*(2), 5–20.

Billon-Galland, A., & Thomson, A. (2018). *European strategic autonomy: Stop talking, start planning.* European Leadership Network. http://www.jstor.com/stable/resrep22125.

Booth, K., & Wheeler, N. (2007). *The security dilemma: Fear, cooperation and trust in world politics.* Palgrave.

Boulding, K. (1989). *Three faces of power.* Sage.

Brennan, L. (2019, July 19). *One of Europe's smallest states aims to become a space superpower.* World Economic Forum. https://www.weforum.org/agenda/authors/louis-brennan.

Brown, C. (2012). Realism: Rational or reasonable? *International Affairs, 88*(4), 857–866.

Brown, M. E., Coté, O. R., Lynn-Jones, S. M., & Miller, S. E. (Eds.) (2004). *Offense, defense, and war.* MIT Press.

Burger, E., & Bordacchini, G. (2019). *Yearbook on space policy 2017: Security in outer space: Rising stakes for civilian space programmes.* Springer Nature.

Calmes, B., Schummer, L., & Gladysz-Lehmann, B. (2021, December 9). *The space law review: Luxembourg.* The Law Reviews. https://thelawreviews.co.uk/title/the-space-law-review/luxembourg.

Carey, M. (2017, November 16). *Luxembourg was looking for ways to diversify its economy and explore new sectoral avenues.* Interview by Étienne Schneider. Happen. https://gouvernement.lu/fr/actualites/toutes_actualites/interviews/2017/novembre/15-schneider-happen.html.

Carminati, D. (2021). The economics of soft power: Reliance on economic resources and instrumentality in economic gains. *Economic and Political Studies, 10*(2). https://doi.org/10.1080/20954816.2020.1865620.

Carr, E. H. (1939). *The twenty years crisis, 1919 1939: An Introduction to the study of International Relations.* Macmillan.

Chong, A., & Maass, M. (2010). Introduction: The foreign policy power of small states. *Cambridge Review of International Affairs, 23*(3), 381–382.

Cohen, B. J. (2008). *International political economy: An intellectual history.* Princeton University Press.

Dalesio, E. P. (2021, August 27). Luxembourg one of EU's defence laggards, data show. *Luxembourger Wort.* https://luxtimes.lu.

Dario, L. (2020, August 22). Joseph Nye: "Trump is the United States' biggest problem." *Bueonos Aires Times.* https://www.batimes.com.ar/news/world/joseph-nye-trump-is-the-united-states-biggest-problem.phtml.

De Martino, M. (2020). Soft power: Theoretical framework and political foundations. *Przegląd Europejski*, 2020(4), 11–24. DOI: 10.31338/1641-2478pe.4.20.1

de Montluc, B. (2009). The new international political and strategic context for space policies. *Space Policy*, 25(1), 20–28.

de Montluc, B. (2012). What is the state of play in European governance of space policy? *Space Policy*, 28(3), 74–76.

Digester, P. (1992). The fourth face of power. *The Journal of Politics*, 54(4), 977–1007.

Elman, C. (2004). Extending offensive realism: The Louisiana Purchase and America's rise to regional hegemony. *American Political Science Review*, 98, 563–576.

Enrikson, A. K. (2001). A coming "Magnesian" age? Small states, the global system, and the international community. *Geopolitics*, 6(3), 49–86.

European Union (EU). (2021). *Joint declaration on the conference on the future of Europe.* https://ec.europa.eu/info/sites/default/files/en_joint_declaration_on_the_conference_on_the_future_of_europe.pdf.

Fan, Y. (2008). Soft power: Power of attraction or confusion. *Place Branding and Public Diplomacy*, 4(2), 147–158.

Finnemore, M., & Goldstein, J. (2013). Puzzles about power. In M. Finnemore & J. Goldstein (Eds.), *Authority, coercion, and power in international relations, back to basics* (pp. 55–77). Oxford University Press.

Foust, J. (2020, January 16). Luxembourg establishes space industry venture fund. *Space News.* https://spacenews.com/luxembourg-establishes-space-industry-venture-fund/.

Frentz, J. M. (2010). The foreign policy of Luxembourg. In R. Steinmetz & A. Wivel (Eds.), *Small states in Europe* (pp. 131–146). Ashgate Publishing.

Gallarotti, G. M. (2011). Soft power: What it is, why it's important, and the conditions for its effective use. *Journal of Political Power*, 4(1), 25–47.

Georgiou, P. (2008). The concept of power: A critique and an alternative. *Australian Journal of Politics & History*, 23(2), 252–267.

Gerhardstein, N. A. (2020, September 24). *The coming "revolution" in the satellite sector.* Delano. https://delano.lu/article/delano_coming-revolution-satellite-sector?index=0.

Gilpin, R. (1981). *War and change in world politics.* Cambridge University Press.

Glaser, C. L. (2010). *Rational theory of international politics: The logic of competition and cooperation.* Princeton University Press.

Gray, C. S. (2011, April). *Hard power and soft power: The utility of military force as an instrument of policy in the 21st century.* Strategic Studies Institute, US Army War College. http://www.jstor.org/stable/resrep11431.

Grieco, J. M. (1990). *Cooperation among nations: Europe, America, and non-tariff barriers to trade.* Cornell University Press.

GovSat. (2022). *Mission.* https://govsat.lu/about-us/.

Haas, M. (2015). Vulnerable frontier: Militarized competition in outer space. In O. Thränert & M. Zapfe (Eds.), *Strategic trends 2015: Key developments in global affairs* (pp. 63–80). ETH — Center for Security Studies.

Haass, R. N. (2008). The age of nonpolarity: What will follow U.S. dominance. *Foreign Affairs*, 87(3), 44–56.

Harmsen, R., & Högenauer, A. L. (2021). Luxembourg and the European Union. In F. Laursen (Ed.), *The Oxford encyclopedia of European Union politics.* https://oxfordre.com/politics/page/european-union-politics/the-oxford-encyclopedia-of-european-union-politics; https://doi.org/10.1093/acrefore/9780190228637.013.1041

Harris, T. (2018). Foreign policy in an age of austerity: "Soft power", hard choices. *L'Observatoire de la Société Britannique, 20,* 17–36. https://journals.openedition.org/osb/1976

Harvey, D. (2005). *The new imperialism.* Oxford University Press.

Hayden, P. (2005). *Cosmopolitan global politics.* 1st edition. London: Routledge.

Higgins, J. (2007). *Satellite newsgathering.* Focal Press.

Hoerber, T. C. (2009). The European space agency and the European Union: The next step on the road to the stars. *Journal of Contemporary European Research, 5*(3), 405–414.

Hoerber, T. C. (2015). Chaos or consolidation? Post-war space policy in Europe. In T. C. Hoerber & P. Stephenson (Eds.), *European space policy, European integration and the final frontier* (pp. 15–29). Routledge.

Hrozensky, T. (2016). *Space – a soft power tool for Europe? Voices from the space community 78.* European Space Policy Institute. https://espi.or.at/publications/voices-from-the-space-community/category/3-voices-from-the-space-community.

Huberty, M. (2011). *Punching above its weight? A case study of Luxembourg's policy effectiveness in the European Union* [Doctoral dissertation, University of Sussex].

Huntington, S. P. (1996). *The clash of civilizations and the remaking of world order.* Touchstone.

Huntley, W. L. (2007). Smaller state perspectives on the future of space governance. *Astropolitics: The International Journal of Space Politics & Policy, 5*(3), 237–271.

International Centre for Defence Studies, Estonia. (2013). *Estonia's national defence, civil security and public safety: Why, what and how?* https://icds.ee/wp-content/uploads/2013/ICDS%20Policy%20Paper_Dual-use%20R&T_T%20Jermalavicius%20M%20Lellsaar_June%202013.pdf

Jermalavičius, T., & Lellsaar, M. (2013). *Dual-use research and technology (R&T) for Estonia's national defence, civil security and public safety: Why, what and how?* https://icds.ee/wp-content/uploads/2013/ICDS%20Policy%20Paper_Dual-use%20R&T_T%20Jermalavicius%20M%20Lellsaar_June%202013.pdf

Jervis, R. (1978). Cooperation under the security dilemma. *World Politics, 30,* 167–214.

Johnson, D. J., & Levite, A. E. (2003). Summary. In D. J. Johnson & A. E. Levite (Eds.), *Toward fusion of air and space: Surveying developments and assessing choices for small and middle powers* (pp. vii–xvi). RAND Corporation.

Johnson-Freese, J. (2007). *Space as a strategic asset.* Columbia University Press.

Kagan, R. (2008). *The return of history and the end of dreams.* Atlantic Books.

Kahler, M. (1998). Rationality in international relations. *International Organization, 52*(4), 919–941.

Karampekios, N., & Oikonomou, I. (2018). The European arms industry, the European Commission and the preparatory action for security research: Business as usual? In N. Karampekios, I. Oikonomou & E. G. Carayannis (Eds.), *The emergence of EU defense research policy from innovation to militarization* (pp. 181–204). Springer.

Kelin, J. J. (2006). *Space warfare: Strategy, principles and policy.* Routledge.

Kolczyński, P. (2018). The new European Union space policy in order to maintain Europe's position among space leaders, CES Working Papers, ISSN 2067-7693, AlexandruIoanCuza University of Iasi, Centre for European Studies, Iasi, *10*(3), 341–356.

Kolovos, A., & Pilaftsis, K. (2015). European multinational satellite programs. In K. U. Schrogl et al. (Eds.), *Handbook of space security* (pp. 843–866). Springer.

Krige, J. (2007). NASA as an instrument of U.S. foreign policy. In S. J. Dick & R. D. Launius (Eds.), *Societal impact of spaceflight* (pp. 207–218). NASA SP2007–4801.

Lake, D. A., Finnemore, M., & Goldstein, J. (Eds.) (2013). *Authority, coercion, and power in international relations, back to basics* (pp. 55–77). Oxford University Press.

Lambakis, S. J. (2001). *On the edge of Earth: The future of American space power.* University of Kentucky Press.

Lambert, Y. (2021a, February 19). Luxembourg votes to increase military spending. *Luxemburger Wort.* https://www.luxtimes.lu/en/luxembourg/luxembourg-votes-to-increase-military-spending-6033ce4ade135b9236b153ae.

Lambert, Y. (2021b, June 15). Luxembourg to fund NATO space project with millions. *Luxemburger Wort.* https://www.luxtimes.lu/en/luxembourg/luxembourg-to-fund-nato-space-project-with-millions-60c8ae42de135b92366dde6b

Lampton, D. (2008). *The three faces of Chinese power: Might, money, and minds.* University of California Press.

Layton, P. (2021). Fighting artificial intelligence battles. Operational concepts for future AI-enabled wars. *Joint Studies Paper Series No. 4.* Australian Defence College. https://www.academia.edu/48963556/Fighting_Artificial_Intelligence_Battles_Operational_Concepts_for_AI_Enabled_Wars

Lee, G. (2009). A theory of soft power and Korea's soft power strategy. *Korean Journal of Defense Analysis, 21*(2), 205–218.

Lee, S.-W. (2011). The theory and reality of soft power: Practical approaches in East Asia. In S. J. Lee & J. Melissen (Eds.), *Public diplomacy and soft power in East Asia* (pp. 11–32). Palgrave Macmillan.

Lippert, B., Ondarza, N. V., & Perthes, V. (Eds.) (2019). European strategic autonomy: Actors, issues, conflicts of interests. *SWP Research Paper*, 4/2019. Stiftung Wissenschaft und Politik – SWP – Deutsches Institut für Internationale Politik und Sicherheit. https://www.ssoar.info/ssoar/handle/document/62346.

Lobell, S. E. (2017). Structural realism/offensive and defensive realism. *Oxford Research Encyclopedias.* https://doi.org/10.1093/acrefore/9780190846626.013.304.

Long, T. (2017). Small states, great power? Gaining influence through intrinsic, derivative, and collective power. *International Studies Review, 19*, 185–205.

Luxembourg Government (LG). (2017). *Luxembourg defence guidelines for 2025 and beyond.* https://defense.gouvernement.lu/dam-assets/la-defense/luxembourg-defence-guidelines-for-2025-and-beyond.pdf.

Luxembourg Government (LG). (2018). *Successful take-off for the Luxembourg space agency.* Press release, September 12. https://www.investinluxembourg.tw/news/successful-take-off-for-the-luxembourg-space-agency/.

Luxembourg Government (LG). (2019a). *The university of Luxembourg launches interdisciplinary space master.* Press release, February 18. https://www.tradeandinvest.lu/news/the-university-of-luxembourg-launches-interdisciplinary-space-master/.

Luxembourg Government (LG). (2019b). *Euro-Composites will invest €160 million in Luxembourg.* Press release, June 16. https://www.tradeandinvest.lu/news/euro-composites-will-invest-e160-million-in-luxembourg/.

Luxembourg Government (LG). (2020a). *National plan for smart, sustainable and inclusive growth Luxembourg 2020, National Reform Programme of the Grand Duchy of Luxembourg under the European semester 2020.* Luxembourg: Ministry of the Economy.

https://2020-european-semester-national-reform-programme-luxembourg_en.pdf (europa.eu).

Luxembourg Government (LG). (2020b). *Présentation du programme LUXEOSys.* https://gouvernement.lu/dam-assets/documents/actualites/2020/07-juillet/2020-07-13-Presentation-Presse-LUXEOSys.pdf.

Luxembourg Government (LG). (2021a). *Luxembourg launches world's first space resources start-up support programme.* Press release, October 26. https://www.tradeandinvest.lu/news/luxembourg-launches-worlds-first-space-resources-start-up-support-programme/.

Luxembourg Government (LG). (2021b). *Luxembourg-based space fund closes at €120 million.* Press release, November, 3. https://www.luxinnovation.lu/news/luxembourg-based-space-fund-closes-at-e120-million/.

Luxembourg Space Agency (LSA). (2017). *Law of July 20th 2017 on the exploration and use of space resources.* Luxembourg Space Agency (LSA).

Luxembourg Space Agency (LSA). (2019). *Space policy and strategy.* https://space-agency.public.lu/en/agency/mission-vision.html

Luxinnovation. (2021). Luxembourg sets out to conquer space. *Financial Times.* https://www.ft.com/partnercontent/luxinnovation/luxembourg-sets-out-to-conquer-space.html.

Luzin, P. (2013). Outer space as Russia's soft-power tool. *Security Index: A Russian Journal on International Security, 19*(1), 25–29. https://doi.org/10.1080/19934270.2013.757117.

Machuron, C.-L. (2022). *Now launching: The space nation. Interview with Franz Fayot, Minister of the Economy and Minister for Development Cooperation and Humanitarian Affairs of Luxembourg, January 17.* Silicon Luxembourg. https://www.siliconluxembourg.lu/now-launching-the-luxembourg-space-nation/.

Mastanduno, M. (1998). Economics and security in statecraft and scholarship. *International Organization, 52*(4), 825–854.

McClory, J. (2010). *The new persuaders: An international ranking of soft power.* Institute for Government. https://www.instituteforgovernment.org.uk/sites/default/files/publications/The%20new%20persuaders_0.pdf.

Mearsheimer, J. J. (1990). Back to the future: Instability in Europe after the Cold War. *International Security, 15,* 5–57.

Mearsheimer, J. J. (2016). Structural realism. In T. Dunne, M. Kurki & S. Smith (Eds.), *International relations theories: Discipline and diversity* (4th ed., pp. 71–88). Oxford University Press.

Melissen, J., Okano-Heijmans, M., & van Bergeijk, P. A. G. (2011). Economic diplomacy: The issues. *The Hague Journal of Diplomacy, 6*(1), 1–6.

Morgenthau, H. J. (1960). *Politics among nations* (3rd ed.). Alfred A. Knopf.

Nikam, O. (2019, May 23). *Luxembourg Space Forum: Consolidating global space cluster.* Satellite Markets & Research. http://satellitemarkets.com/news-analysis/luxembourg-space-forum-consolidating-global-space-cluster.

Nye, J. S. (1990). *Bound to lead: The changing nature of American power.* Basic Books.

Nye, J. S. (2002). *The paradox of American power: Why the world's only superpower can't go it alone.* Oxford University Press.

Nye, J. S. (2004). *Soft power: The means to success in world politics.* Public Affairs.

Nye, J. S. (2006, February 23). *Think again: Soft power.* Foreign Policy. https://foreignpolicy.com/2006/02/23/think-again-soft-power/.

Nye, J. S. (2014). The information revolution and soft power. *Current History, 113*(759), 19–22. https://dash.harvard.edu/handle/1/11738398.

Nye, J. S. (2017). Soft power: The origins and political progress of a concept. *Palgrave Communications, 3,* 17008. https://doi.org./10.1057/palcomms.2017.8.

OECD. (2019). *The space economy in figures: How space contributes to the global economy.* OECD Publishing. https://doi.org/10.1787/c5996201-en.

Oglesby, K. (2021a, October 11). Companies to help Luxembourg run costly satellite. *Luxembourg Times.* https://www.luxtimes.lu/en/luxembourg/companies-to-help-luxembourg-run-costly-satellite-61645e52de135b9236c94c0e.

Oglesby, K. (2021b, November 4). US pull-out from Afghanistan hurt SES revenue forecast. *Luxembourg Times.* https://www.luxtimes.lu/en/luxembourg/us-pull-out-from-afghanistan-hurt-ses-revenue-forecast-6183b64ede135b923618d8a5.

Ombredane, E. (2021, December 7). *SES supports US army's modern satellite warfare.* Delano. https://delano.lu/article/ses-supports-us-army-s-modern-.

Paikowsky, D., & Ben Israel, I. (2009). Science and technology for national development: The case of Israel's space program. *Acta Astronautica, 65*(9–10), 1462–1470.

Palacio, A. (2020). Prologue: European defence and security at a time of global mutation. In M. Ramírez & J. Biziewski (Eds.), *Security and defence in Europe* (pp. vii–xvii). Springer.

Papadimitriou, A., Antoni, N., Adriaensen, M., & Giannopapa, C. (2019). Perspective on space and security policy, programmes and governance in Europe. *Elsevier Open Access.* https://www.sciencedirect.com/science/article/pii/S0094576518303485.

Pasco, X. (2019, October 5). A new French Space Command. *Space Alert, VII*(4), 1–13. https://www.orfonline.org/research/space-alert-volume-vii-issue-4-56195/.

Peter, N. (2009, April 2009). *Space power and Europe in the 21st century.* ESPI Perspectives Nr 21.

Pollpeter, K. (2008). *Building for the future: China's progress in space technology during the tenth 5-year plan and the U.S. response.* Strategic Studies Institute, U.S. Army War College.

Price Waterhouse Coopers (PWC). (2017, August 1). *Leading the space odyssey towards new frontiers.* https://www.pwc.lu/en/commercial-companies/docs/pwc-space.pdf.

Pultarova, T. (2017, November 7). Luxembourg eyes Earth-observation satellite for military and government. *Space News.* https://spacenews.com/luxembourg-eyes-earth-observation-satellite-for-military-and-government/.

Rifkin, J. (2011). *The third industrial revolution. How lateral power is transforming energy, the economy, and the world.* Palgrave Macmillan.

Riker, W. H. (1964). Some ambiguities in the notion of power. *American Political Science Review, 58*(2), 341–349.

Robinson, D. K. R., & Mazzucato, M. (2019). The evolution of mission-oriented policies: Exploring changing market creating policies in the US and European space sector. *Research Policy, 48*(4), 936–948.

Rothman, S. B. (2011). Revising the soft power concept: What are the means and mechanisms of soft power? *Journal of Political Power, 4*(1), 49–64.

Rothstein, R. L. (1968). *Alliances and small powers.* Columbia University Press.

Sagath, D., Papadimitriou, A., Adria Ensen, M., & Giannopapa, C. (2018). Space strategy and governance of ESA small member states. *Acta Astronautica, 142*(1), 112–120.

Saperstein, H. T., & Cera, L. (2021, January 7). *Thailand's smaller-state space power amid great-power competition in the space domain.* Asia Centre. https://centreasia.eu/thailands-smaller-state-space-power-amid-great-power-competition-in-the-space-domain/.

Schnuer, C. (2020a, October 27). *Parliamentary majority blocks satellite inquiry.* Delano. https://delano.lu/article/delano_parliamentary-majority-blocks-satellite-inquiry.

Schnuer, C. (2020b, November 9). *Kleos launches scouting mission satellites.* Delano. https://delano.lu/article/delano_kleos-launches-scouting-mission-satellites.

Schnuer, C. (2020c, November 11). *Budget for military satellite to undergo audit.* Delano. https://delano.lu/article/delano_budget-military-satellite-undergo-audit.

Schnuer, C. (2021a, June 24). *Space new frontier for Lux-US military cooperation.* Delano. https://delano.lu/article/delano_space-new-frontier-lux-us-military-cooperation.

Schnuer, C. (2021b, October 12). *Military satellite on track after budget snafu.* Delano. https://delano.lu/article/military-satellite-on-track-af.

Schweller, R. L. (1994). Bandwagoning for profit: Bringing the revisionist state back in. *International Security, 19*(1), 72–107.

Shadunts, A. (2016, October 28). *The rational actor assumption in structural realism.* E-International Relations. https://www.e-ir.info/2016/10/28/the-rational-actor-assumption-in-structural-realism.

Sheehan, M. (2007). *The international politics of space.* Routledge.

Sheetz, M. (2017, November 11). *The space industry is now 2% of Luxembourg's GDP, Deputy Prime Minister Etienne Schneider says. Interview with Deputy Prime Minister Etienne Schneider.* CNBC. https://www.cnbc.com/2017/11/11/etienne-schneider-the-space-industry-is-now-2-percent-of-luxembourgs-gdp.html.

Société Européenne des Satellites (SES). (2018a). *Investor presentation 2018.* https://www.ses.com/sites/default/files/2018-09/180831_IR%20presentation_Sep%2018_FINAL_WEB.pdf

Société Européenne des Satellites (SES). (2018b). *SES provides managed services for Galileo.* Press release, May 30. https://www.ses.com/press-release/ses-provides-managed-services-galileo.

Société Européenne des Satellites (SES). (2019). *Broadening digital access and services.* Press release, August 7. https://www.ses.com/case-study/burkina-faso-government.

Société Européenne des Satellites (SES). (2021a). *Investor presentation 2021.* https://www.ses.com/sites/default/files/2021-11/Roadshow_Presentation_November2021.pdf.

Société Européenne des Satellites (SES). (2021b). *SES government solutions releases new unified operational network.* Press release, December 6. https://www.ses.com/press-release/ses-government-solutions-releases-new-unified-operational-network.

Société Européenne des Satellites (SES). (2022). *Our history highlights.* https://www.ses.com/about-us/our-history-highlights.

SpaceWatch.Global (SWG). (2020). Étienne Schneider, Architect of Luxembourg's space sector, to step down in February 2020. Spacewatch. https://spacewatch.global/2020/01/etienne-schneider-architect-of-luxembourgs-space-sector-to-step-down-in-february-2020/.

Spire Global, Inc. (Spire). (2018). *World's largest space to cloud data company provides space-based data to one of Japan's largest companies.* Press release, December 10. https://ir.spire.com/news-events/press-releases/detail/27/japans-itochu-taps-spire-for-space-based-data.

Spire Global, Inc. (Spire). (2019). *Kleos and Spire partner on safety at Sea.* Press release, August, 6. https://ir.spire.com/news-events/press-releases/detail/19/kleos-and-spire-partner-on-safety-at-sea.

Spire Global, Inc. (Spire. (2022). *Spire and dragonfly aerospace announce partnership to support Australian office of national intelligence.* Press release, January 21. https://ir.spire.com/news-events/press-releases/detail/89/spire-and-dragonfly-aerospace-announce-partnership-to.

Stiles, K. W. (2018). *Trust and hedging in international relations.* The University of Michigan Press.

Tellis, A. J., Bially, J., Layne, C., & McPherson, M. (2000). *Measuring national power in the postindustrial age.* RAND Corporation. https://www.rand.org/pubs/monograph_reports/MR1110.html.

Telò, M. (2006). *Europe: A civilian power? European Union, Global governance, world order.* Palgrave.

Thorhallsson, B. (2011). Domestic buffer versus external shelter: Viability of small states in the new globalised economy. *European Political Science, 10*(3), 324–336.

Thucydides. (1972). *History of the Peloponnesian War* (R. Warner, Trans.). Penguin Classics. (Original work published 431 BCE).

Treverton, G. F., & Jones, S. G. (2005). *Measuring national power.* Rand National Security Research Division Conference Proceedings. RAND Corporation. https://www.rand.org/pubs/conf_proceedings/CF215.html.

Trouillez, M.-H. (2019, July-August). Luxembourg in the conquest of space. Interview with Étienne Schneider, Deputy Prime Minister, Minister of the Economy. *Merkur.* https://meco.gouvernement.lu/en/actualites.gouvernement%2Ben%2Bactualites%2Btoutes_actualites%2Binterviews%2B2019%2B07-juillet%2B05-schneider-merkur.html.

Tziortzis, S. (2020). *Cyprus activities in space.* Presentation at the 1st virtual Excelsior International Technical Workshop, July 15. http://libraryaplos.com/bitstream/123456789/4812/1/20200715_EXCELSIOR_WP9_CYPRUSACTIVITIESINSPACE_V1_PU.pdf.

Van Evera, S. (1998). Offense, defense, and the causes of war. *International Security, 22*(4), 5–43.

Venet, C. (2011). The economic dimension. In C. Brünner & A. Soucek (Eds.), *Outer space in society, politics and law* (pp. 55–72). Springer.

Walt, S. M. (1987). *The origin of alliances.* Cornell University Press.

Waltz, K. N. (1979). *Theory of international politics.* Addison-Wesley.

Weber, M. (2012). *The theory of social and economic organization* (A. M. Henderson & T. Parsons, Trans.) (T. Parsons, Ed.). Martino Fine Books.

White, T. H. (1978). *In search of history: A personal adventure.* Harper & Row.

Wiberg, H. (1987). The security of small nations: Challenges and defences. *Journal of Peace Research, 24*(4), 339–363.

Wivel, A. (2010). From small state to smart state: Devising a strategy for influence in the European Union. In R. Steinmetz & A. Wivel (Eds.), *Small states in Europe: Challenges and opportunities* (pp. 15–30). Ashgate.

Wrong, D. (2017). *Power: Its forms, bases and uses.* Routledge.

Part III
The Global Dimension

9 From Space Situational Awareness to Space Domain Awareness

Examining Rhetorical and Substantive Transitions in the U.S. Approach to Space Security

Mariel Borowitz

Introduction: U.S. Military Space Activity

For years, the U.S. military has maintained the most advanced space surveillance system in the world, monitoring the status of human-made objects in space, and providing data and information to spacecraft operators around the world to help avoid collisions in space. In 2019, the United States re-established USSPACECOM and created the U.S. Space Force. In concert with these organizational changes, the U.S. military announced in October 2019 that it would adopt the term "SDA," in place of the previously used – and broadly accepted – term "SSA" (Erwin 2019). According to the memo announcing the change, the new terminology was designed to reflect the shift in focus from space as a benign environment to space as a warfighting domain. This chapter examines how both the rhetoric and the substantive approach to SSA/SDA has shifted in recent years, treating this as an illustrative case study to examine whether there is a trend toward militarization in U.S. space activities.

The U.S. military has been involved in space activities since before the first satellite was launched. In its first statement of national policy for outer space, contained in a 1955 U.S. National Security Council document, the United States acknowledged the practical and strategic importance of space. It endorsed the U.S. goal of launching a small satellite as part of the International Geophysical Year but made clear that this endeavor must not slow progress on efforts to develop a more advanced reconnaissance satellite (NSC 1955). In the years following, parallel to the civil efforts undertaken by NASA, the U.S. military developed numerous types of space technologies that would support their mission – reconnaissance; early warning; communication; and positioning, navigation, and timing satellites. The U.S. military has continued to develop and refine these systems over the last 60 years.

In addition to satellites designed to support and enhance forces on the ground, the United States also developed and tested anti-satellite weapons, a capability designed to deny freedom of action in space to potential adversaries. By the late

DOI: 10.4324/9781003230670-13

1950s, the United States had already tested multiple anti-satellite capabilities. U.S. testing of anti-satellite weapons ended in the 1980s, and none were ever used against an adversary's spacecraft. In 2008, the United States demonstrated that it still has the capability to destroy a satellite in orbit through Operation Burnt Frost, which used a modified ballistic missile to destroy a de-orbiting U.S. military satellite, ostensibly to avoid the possibility of the spacecraft surviving re-entry and causing damage on the ground (Weeden and Samson 2021).

To support these activities, and to gain insight into the activities of others, the United States developed the ability to detect and track satellites, allowing officials to understand where these objects are and where they will be in the near future. This SSA capability is particularly interesting, because it has both a security and a safety aspect. High-quality SSA is required to carry out attacks on space assets – providing key targeting information. SSA is also essential to attribution, which improves deterrence – adversaries are more likely to carry out attacks if their actions cannot be attributed, and less likely to attack if attribution is likely. However, the value of SSA data goes beyond military use. This data has become increasingly important to all satellite operators in order to identify and avoid accidental collisions. Currently, the U.S. military makes SSA data available to satellite operators around the world, and while some other nations and commercial entities operate their own SSA systems, the U.S. military is widely recognized as the most significant global provider of SSA data and services.

In 2019, U.S. military space activity underwent a significant organizational change, with the creation of USSPACECOM in August 2019, and the creation of the U.S. Space Force in December 2019. USSPACECOM is one of 11 combatant commands. It conducts operations in, from, and to space. The U.S. Space Force is a new military service, similar to the Army or Navy. U.S. Space Force organizes, trains, and equips space forces. Its mission includes "developing Guardians, acquiring military space systems, maturing the military doctrine for spacepower, and organizing space forces to present to our Combatant Commands."(USSF 2022) The creation of these two new organizations has the potential to change the trajectory of U.S. military space activity, and some fear that it may lead to increased militarization of space. Perhaps, a signal of such a change, in the midst of this organizational transition, the U.S. military announced that it would adopt the term "SDA," replacing the term "SSA," in recognition that space is considered a domain of warfare, just like land, air, and sea.

This chapter examines the trajectory of U.S. SSA activities, and particularly the changes that have occurred following the transition to the term SDA, with the goal of identifying whether this shift in terminology represents a shift toward increased militarization of space within the United States. I examine how the rhetoric related to SSA/SDA has changed within the U.S. military, and to what extent this change is reflected in the substantive approach to SSA/SDA activities and issues. Does this change in terminology represent a change in the way the U.S. military views or approaches the issue of SSA? Is it a sign of increased militarization of space?

Defining Militarization

Understanding whether "militarization of space" is accelerating requires that we first understand the meaning of this term. Many authors define militarization of space simply as the military use of space, and most acknowledge that by this definition, space has been militarized since the beginning of the space age. For example, Rosas (1983) noted that already at that time, space was an important theater for military activities, with an estimated 75% of satellites in orbit at the time performing military missions (Rosas 1983: 357). In "The Myths of Space Militarization," Aldridge (1984) details the early involvement of the U.S. military in space activities, showing that military use of space is nothing new (Aldridge Jr 1987).

Other authors, such as Freese and Burbach (2019), use the term "weaponization of space," focusing on offensive actions in, and from, space (Johnson-Freese and Burbach 2019). As noted above, offensive space weapons have been tested, but not deployed in conflict. Peoples (2020) argues that the distinction between the terms militarization and weaponization is not significant because many space capabilities are dual use. Even if designed as a passive system, it may be used in an offensive way. A common example is the Space Shuttle, which had the capability to rendezvous with spacecraft. While designed to allow satellite repair, the same capability could allow destruction of an enemy satellite (Peoples 2010). The intent of the organization controlling the space asset is the key variable – and that variable cannot be independently observed. Defrieze (2014) argues that a space weapon cannot be defined by its physical properties or functions but rather must be defined by what it is used to do. He argues that the focus must be on the behaviors we find acceptable or unacceptable (DeFrieze 2014). However, with this definition, you could only identify "weaponization" after an offensive act had occurred. While it is surely important to have a shared understanding of acceptable and unacceptable behaviors in space, it is still prudent to also pay attention to technological development, particularly those that offer offensive capabilities – even if latent.

It is also possible to understand these terms not as dichotomous states but as a spectrum along which development may occur. This could be measured as the volume or proportion of space activity undertaken by the military – how many space objects, or what proportion of space objects, are owned or operated by the military? How are military investments in space changing? What is the rhetoric that surrounds the development or use of these technologies? Drawing from the weaponization literature, we may consider trends in offensive versus defensive space developments in technology and/or doctrine. Taking this approach, it is possible to consider whether, and in what ways, militarization of space may be increasing or decreasing over time.

Theoretical Framework

Militarization of space can be further understood by examining the concept through the lens of traditional international relations theory, particularly the dichotomy between realist and liberal approaches to the development of space. Grondin (2009) argues that the difference between the two camps revolves primarily around the inevitability of the weaponization of space.

Realist theory posits that nations work to achieve their own self-interest within an anarchic world system, such that power and security are central to understanding world events. Realists tend to view conflict in space, and hence militarization and weaponization, as inevitable. Moltz (2020) explains that in the realm of space, realism suggests that leading spacepowers will continue military developments in space in an effort to gain geopolitical advantages (Moltz 2020: 24). Grondin (2009) notes that this view has been reflected in multiple U.S. Space Policy documents.

By contrast, liberalism suggests that the anarchic global system can be moderated through international institutions, organizations, and norms of behavior. Individuals with this approach reject the idea that space assets can only be protected through the development of space weapons (Moltz 2020: 36). Viewed through this lens, weaponization of space can be avoided through treaties and other international agreements. As DeFrieze (2014: 110) states, "Space is an international common and is thus easier to protect through international cooperation." Indeed, the international space community has succeeded in negotiating multiple important treaties governing the use of outer space. Treaties have even addressed some space weapons, with the Partial Test Ban Treaty, for example, banning nuclear testing in outer space. Liberal theory also highlights the importance of economic and societal interdependence as a force that decreases the likelihood of conflict among states. In the space domain, there is high interdependence due to the nature of the space environment; creation of debris or other degradation of the space environment has the potential to harm many types of users – civil, military, and commercial – from many nations (Hansel 2010).

A third view, focused on domestic politics and bureaucratic theory, is also worth considering, particularly with the recent re-organization of U.S. military space. Johnson-Freese and Burbach (2019) suggest that the Space Force, in an effort to justify its existence and make its place within the military structure, must present itself as a fighting force. As such, it will emphasize threats to space assets and the offensive capabilities needed to defeat them (Johnson-Freese and Burbach 2019). Some fear that this aggressive rhetoric could lead other nations to perceive the United States as a greater threat, leading those nations to also step up their own offensive space capabilities. In this view, the emphasis on threats in space could become a self-fulfilling prophecy.

As noted above, SSA capabilities can be used for offensive or defensive military capabilities, as well as for safety of flight relevant for all space operators. When prioritizing SSA development, realists would emphasize the importance of offensive capabilities, ensuring the United States has the high-quality SSA needed for identifying and targeting enemy spacecraft. This includes not only improving data collection systems and tracking algorithms but also intelligence that allows the United States to understand which satellites may pose a threat.

Liberal thinkers may instead seek to improve SSA capabilities – and security – by engaging with international partners. Sharing SSA data and creating a common understanding of the space environment can help to improve deterrence and decrease the likelihood that attacks on space assets will occur. Coordination with global commercial entities, with an emphasis on safety and sustainability of the space environment, would also be a priority in this view. Security could be further increased by engaging not only with allies but also with potential adversaries. Such engagement could include SSA data sharing or agreements on appropriate behavior in space.

Building on the concepts put forward by Johnson-Freese and Burbach (2019), this paper also considers the possibility that changes in SSA are primarily rhetorical, driven by domestic politics, particularly associated with the creation of USS-PACECOM and the U.S. Space Force. In this case, we may find that changes in terminology and associated rhetoric are not accompanied by significant substantive changes. However, it is also possible that changes that were originally only rhetorical may result in substantive changes, due to the dynamics described above.

The following section provides a detailed examination of the development of SSA capabilities in the United States. Technical developments and improvements and organizational challenges are examined, as well as the rhetoric surrounding these activities in speeches and key policy documents.

Developments and Trends in SSA

Early Space Surveillance Efforts

Plans to track objects in space – an activity originally referred to simply as "space surveillance" or "space tracking" – were put in place in the mid-1950s, in preparation for launches expected during the International Geophysical Year. The earliest efforts were undertaken by universities and volunteer amateur astronomers, supported with funding from the U.S. Air Force. The Naval Research Laboratory also developed a complementary electronic detection system (Sturdevant 2008: 6–7).

Military-focused efforts accelerated soon after. The Naval Space Surveillance system began more systematic efforts at tracking in 1958. That same year, the Air Force issued General Operational Requirement 170, which called for a space tracking and control system, followed by anti-satellite weapons. In 1959, officials at the North American Air Defense Command [North American Aerospace Defense Command (NORAD)] advocated for the creation of a catalog of all objects in space that could be used in efforts to develop active defense capabilities (Sturdevant 2008: 10). These early efforts clearly emphasized the importance of space tracking for offensive space capabilities.

In 1961, NASA and the Department of Defense (DoD) agreed to provide data from their respective sensor systems to a centralized data collection and cataloging center within the NORAD. The DoD removed data on sensitive assets, such as the newly launched Corona reconnaissance satellite, before providing the catalog to NASA, which further disseminated that information to others. By 1963, the DoD developed initial capabilities to estimate the operational mission of objects put into space by other nations (Sturdevant 2008: 12). In these actions, we see the military increasing secrecy and improving capabilities for military-specific applications of the data, while still acknowledging the importance of the space safety mission through cooperation with civil entities.

In following years, the accuracy of the tracking system was improved, both to allow for prediction of U.S. spacecraft re-entry points, and to support operation of the country's first anti-satellite system. In 1967, the Space Defense Center was created within Cheyenne Mountain. Despite its seeming importance, space tracking capabilities were not provided by dedicated systems – the sensors were operated by a variety of organizations for various purposes. The first dedicated space surveillance radar was not added to the system until 1969. Over the next decade, additional tracking resources were added to the network, largely in response to Soviet anti-satellite weapon testing and U.S. anti-satellite weapon development (Sturdevant 2008: 12–15).

In 1982, the Air Force established the Air Force Space Command focused on space operations (USAF 2022). That same year, it carried out a space surveillance architecture study to determine space surveillance requirements and identify sensors needed to meet those requirements. This process identified the need for a space-based space surveillance system. A 1986 review of the Space-Based Space Surveillance System program carried out by the General Accounting Office (GAO) stated that the purpose of space surveillance was "to detect, track, identify, and assess space objects of all kinds, especially satellites and antisatellite interceptors." This included determining both the location and the mission of objects launched by other nations. GAO explained, "With this information, actions can be taken to protect U.S. satellites from attack or initiate attacks against enemy satellites should the need arise" (GAO 1986: 2). A 1987 report reiterated that the Space Defense mission included the use of space surveillance capabilities to support U.S. anti-satellite weapons and warn of attacks on U.S. satellites (GAO 1987: 1). Once again, these documents highlight the emphasis on using SSA data for national security purposes, particularly detecting and responding to potential threats.

Throughout the early space age, the DoD had continued its coordination with NASA. Following the Challenger accident in 1986, the DoD provided warnings to NASA when tracked objects would come close to the Space Shuttle (GAO 1990: 4). Despite these activities, DoD officials explained to General Accountability Office investigators that as of 1990, they did not have the capability to detect and identify all debris, and they had not approached the issue of space debris from a collision-avoidance perspective (GAO 1990: 15). This suggests that while the military was willing to share data with civil entities, this was largely a by-product of their core goal, which was squarely focused on the national security relevance of space surveillance.

Increasing Importance of Space and Shift to SSA

Throughout the 1990s and early 2000s, the use of space – by the U.S. military as well as many other users – increased significantly. Operation Desert Storm, in 1991, was considered by many "the first space war." Commercial communications and remote-sensing satellites were proliferating. It was clear that space assets were becoming increasingly important for military operations and for the global economy. Up until this time, spacecraft tracking had been largely a military activity, carried out by the military primarily for military purposes, with the exception of some coordination with NASA, as needed. However, the two realizations mentioned earlier – first, that space assets were increasingly important to military activity and also inherently vulnerable, and second, that growing civil and commercial activity required increased attention and support – led to significant developments in military-focused space surveillance capabilities as well as those related to civil and commercial space safety needs in the late 1990s and early 2000s.

In 1997, the USSPACECOM, which had been established in 1985, published its "Vision for 2020." This document acknowledged growing global capabilities in space. The authors argued that the United States "may evolve into the guardian of space commerce – similar to the historical example of navies protecting sea commerce (USSPACECOM 1997)." It also noted that increased military dependence on space capabilities may lead to increased vulnerabilities, and that given their importance, it was unrealistic to expect that military space systems will never become targets. The document called space the "fourth medium of warfare" along with land, sea, and air and argued that U.S. space forces were needed to "protect military and commercial national interests and investments in the space medium" (USSPACECOM 1997: 10). It called for the United States to "dominate space" (USSPACECOM 1997: 10) and put forward the concept of space control. The report noted that this required surveillance of space, with real-time, precise data (USSPACECOM 1997: 10). A 1997 GAO report stated that the DoD recognized that its space surveillance network could not adequately deal with future national security threats, and that the growing commercial space sector would result in increased requests for surveillance support (GAO 1997: 12).

The USSPACECOM Long Range Plan, released in 1998, focused on the steps necessary to implement the USSPACECOM *Vision for 2020*. It called for "near real-time Space Situational Awareness" – the first use of that term in official documentation – and called it the "foundation for space superiority" (USSPACECOM 1998: 28). This required the capability to characterize high-interest objects – using imagery, electronic intelligence, and other information to identify potential threats. It also called for an improvement in detection and tracking needed to support crewed spacecraft as well as to employ weapons. Finally, the report called for the creation of a more accurate catalog of space objects that would be "shared with all nations" and organizations that own or operate space systems (USSPACECOM 1998: 29). While centralized surveillance of space by USSPACECOM was seen as critical, the organization would also make use of data from a variety of sources. The report suggested that allied or foreign surveillance capabilities could be integrated into an international operations center, with NASA, Canada, the European Space Agency, and the Russian Space Agency identified as potential partners (USSPACECOM 1998: 28–31).

SSA capabilities improved somewhat in the following years, as the United States experimented with the use of space-based SSA and advanced its analysis capabilities to produce a "high-accuracy" catalog. Using "special perturbation" algorithms, this catalog was several times more accurate than the general catalog and was provided only to selected users on a case-by-case basis (Sturdevant 2008: 16). Still, in 2001, two major reports emphasized the need to further improve SSA capabilities. The Report of the Commission to Assess United States National Security Space Management and Organization (typically referred to as the "Rumsfeld Commission" because it was chaired by Donald Rumsfeld) warned of the potential for a surprise attack in space –a "space Pearl Harbor" (Rumsfeld, Andrews et al. 2001: 22). It noted that SSA is the key capability needed to avoid such a scenario (Rumsfeld, Andrews et al. 2001: 31). Similarly, the 2001 Quadrennial Defense Review stated that achieving the space control mission required improvements in SSA (DOD 2001: 45). The military was actively developing such missions – in 2003, it tested the XSS-10 satellite, which was used to approach satellites for inspection, sending imagery that could be used to determine threats (Sturdevant 2008: 18).

The United States made extensive use of space assets for intelligence gathering, targeting, and weapons guidance during the 2003 Iraq War, further illustrating the growing importance of space assets to the military (Posen 2003). In 2004, the Air Force released its first doctrine publication on counterspace operations. Counterspace operations include both the defensive and offensive elements necessary to ensure freedom of action in space for the United States as well as denying that capability to adversaries. The document stated that SSA was the foundation of these capabilities and dedicated a full chapter (one of six in the document) to SSA, clarifying that SSA includes not just space surveillance but also detailed reconnaissance of specific space assets, collection, and processing of intelligence data on space systems, and monitoring of the space environment (i.e. space weather). The document stated that SSA is needed to understand and respond to threats to U.S. space systems and target and engage enemy systems, and while it is critical for all space activities, its most stringent requirements are derived from counterspace operations (USAF 2004: 2).

As the United States continued to push forward in developing military space capabilities, some advocated for a cooperative approach. In 2002, China and Russia submitted a joint working paper on the "Prevention of the Deployment of Weapons in Outer Space" in the United Nations Conference on Disarmament. The Chinese representative warned that outer space was faced with the danger of weaponization and an arms race. The United States remained opposed to such efforts, arguing that the existing outer space regime was sufficient (UN 2002). This position was further codified in the 2006 U.S. National Space Policy, which stated, "The United States will oppose the development of new legal regimes or other restrictions that seek to prohibit or limit U.S. access to or use of space" (United States 2006: 2).

In 2007, China conducted a kinetic anti-satellite test, using a ground-based missile to destroy its own aging weather satellite in low Earth orbit. This action significantly increased the amount of debris in orbit and focused public and Congressional attention on threats to satellites. Critics argued that efforts to improve SSA were not keeping up with this growing risk (Singer 2007). The former chairman of the House Armed Services committee called on the military to improve its SSA capabilities, noting, "we are limited in what we can do in space if we don't know what's going on up there" (Everett 2007). Other experts argued that the most effective response would be to lay out rules of the road for responsible behavior in space (Krepon and Katz-Hyman 2007).

One year after the Chinese anti-satellite test, the United States announced that it would use a modified missile defense interceptor to destroy its own malfunctioning reconnaissance satellite. The United States stated that this was necessary to avoid contamination that could result if the satellite continued its uncontrolled re-entry, since it carried a large amount of hazardous hydrazine fuel that would be likely to survive re-entry. Despite this explanation, many interpreted this action as an anti-satellite test carried out in response to the Chinese actions in 2007. Regardless of the underlying motivation, the United States undertook the activity in a much different fashion, providing advanced warning to the international community and designing the interception to minimize the creation of long-lasting debris (Johnson 2021). SSA information was critical to this activity, providing the precise position of the satellite, projecting its path, and detecting and tracking debris resulting from the interception (Raymond 2007).

Growing Civil Use of Space and Need for SSA

As seen above, throughout the late 1990s and 2000s, U.S. military use of space – including the development of offensive and defensive space capabilities – had increased significantly. Military leaders largely focused on SSA as an enabler of military missions – identifying threats, controlling space, and protecting space assets. But this data had value for space safety, as well. Civil and commercial space activities had increased rapidly around the globe, and in 2002, the U.S. GAO noted that government, commercial, and foreign entities relied almost exclusively on the United States space surveillance network data (provided via NASA) to reduce the risk of space collisions (GAO 2002: 1).

This became a source of concern for the military, which saw the existing system as a potential risk to national security. Space Command officials noted that NASA did not verify the identity of the users accessing SSA data on its website and argued that users with access to this data could attempt to damage or jam satellites or move military assets to avoid detection (GAO 2002: 1). Subsequently, the Air Force Space Command proposed a pilot study to place a Federally Funded Research and Development Center in charge of data dissemination, ending the long-standing arrangement with NASA. Officials argued that this move would allow them to better "maintain control over processes and data dissemination" and "determine if providing support to a foreign entity was in the best interests of national security" (GAO 2002: 3).

In 2003, Congress authorized the DoD[1] to undertake such a pilot program to provide satellite tracking support to entities outside the U.S. government (Congress 2003). The Commercial and Foreign Entities (CFE) pilot program resulted in the development of the space track website, which provided registered users with access to basic information about objects in space. (The high accuracy catalog was not made publicly available.) By 2005, the NASA website had ceased operations, and space-track had over 16,000 registered users (Sturdevant 2008: 19).

Just as China's 2007 anti-satellite test, and the resulting debris, raised awareness of the threat to military satellites, it also led to calls for improved services to the global civil and commercial sector. The CEO of Intelat, a major commercial communications satellite operator, called for increased funding for the U.S. SSA tracking program and improvements to the CFE program. He noted that the data currently provided under the program was not sufficient for independent collision assessments (McGlade 2007). The shortcomings of the system were dramatically illustrated with the 2009 collision of an Iridium commercial communications satellite and a defunct Russian communications satellite, which created thousands of pieces of debris. Until this point, the United States had been focusing its tracking and analysis on U.S. military assets. Following the event, it began conducting analyses for all 800 active satellites in orbit at the time (Shalal-Esa 2009).

International advancements in SSA also challenged the United States to consider the views of a broader range of stakeholders. While the United States had routinely removed its own classified satellites from the publicly provided catalog, it had not removed those of allies. In 2007, when France developed its own independent catalog of space objects, it threatened to publish information on classified U.S. satellites, unless its own military systems were removed from the U.S. database (Space News Editor 2007; the United States ultimately obliged, removing the objects in question). In 2008, the European Space Agency began work on a space surveillance network that would tie together assets from a variety of European nations, the first step toward developing an independent capability for Europe (De Selding 2008). Commercial industry began to take independent action as well, forming the Space Data Association in 2009 to share positional data among commercial satellite operators to enable commercially-relevant SSA (SDA 2022).

In 2008, General Chilton, commander of the U.S. Strategic Command, acknowledged that SSA crosses civil, military, and commercial boundaries and suggested that trans-Atlantic cooperation should be considered (De Selding 2008). Similarly, Lt. General Shelton, commander of U.S. Strategic Command's Joint Functional Component Command for Space, argued that the operators of military, civil, commercial, and allied satellites all need to share more information with one another (Brinton 2008). The military's CFE pilot program was re-named the SSA Data Sharing Program and made permanent (Chow 2011: 2). Driven by increasing foreign and commercial SSA capabilities and demand for data, as well as tightening budgets, in mid-2009, U.S. Air Force Space Command demonstrated its ability to incorporate data from non-Air Force satellite operators into its SSA system (Singer 2009).

This increased engagement with international and commercial partners also aligned with the incoming Obama administration's more cooperative approach to space issues, illustrated in the National Space Policy released in 2010. This document called out the importance of using SSA information "from commercial, civil, and national security sources" to help ensure the "long-term sustainability of the space environment" (United States 2010: 7). The 2011 National Space Security Strategy continued this trend, arguing that the United States should use its leadership in SSA to "foster cooperative SSA relationships, support safe space operations, and protect U.S. and allied space capabilities and operations" (United States January 2011: 6). That same year, the deputy secretary of defense for space policy stated that the United States should negotiate orbital data sharing agreements with other nations as part of its effort to ensure space safety and suggested that the United States could provide space surveillance data as a global service, similar to that provided by the U.S. GPS system (Space News Staff 2011).

Throughout the 2010s, military discussion of SSA continued to acknowledge the importance of international and commercial activities in space and reflected the importance of both space safety and national security concerns. For example, the 2010 Quadrennial Defense Review stated that the DoD would work to promote spaceflight safety as well as use SSA data to enhance the ability to attribute actions in space (DOD 2010: 33).

In practice, the DoD struggled to complete major technological improvements to its SSA system, including the deployment of the Space Fence monitoring system and the Joint Space Operations Center Mission System needed to enable advanced analysis, both of which were significantly over budget and behind schedule (GAO 2011). Commercial users of the military SSA data complained that the quality of data and warnings were insufficient, noting that numerous conjunction warnings received by the military were determined, on closer analysis, to be false alarms, and some true close approaches did not generate warnings from the military (Ferster 2012). Independent international and commercial SSA efforts continued to proliferate, in part to address perceived shortcomings in data provided by the United States.

Risks in space continued to grow, with increased launches and a rise in debris, as well as potentially threatening actions. In 2015, leaders of the communications satellite company, Intelsat, stated that a Russian satellite had maneuvered into a position between two of its satellites, at a distance small enough to pose a risk in the view of company leaders. Russian satellite operators did not respond to initial attempts to communicate by Intelsat or the U.S. DoD (Gruss 2015). Russian officials later denied its actions were aggressive and noted that it had not violated any rules (Peter 2015). In subsequent years, Russia continued to maneuver this satellite near a variety of other commercial and foreign satellites (Roberts 2021). Beginning in 2016, a Chinese satellite began to carry out similar maneuvers, although it only visited other Chinese spacecraft (Roberts 2021). Data released by Russia in 2019 showed that U.S. satellites were undertaking similar behavior, approaching Russian and other foreign satellites beginning in 2016 (Hitchens 2019).

Calls for Civil SSA

Despite the change in rhetoric and the efforts on the part of the DoD to pursue cooperative agreements with international and commercial entities, some experts warned that a military-based SSA system serving national security, civil, commercial, and foreign users was unlikely to succeed. Brian Weeden, a technical advisor at the Secure World Foundation, argued that the U.S. military's decision to conduct conjunction analyses for all operational space objects was primarily driven by a desire to retain control over SSA data. Taking on this new task was preferable to sharing data of sufficient quality to allow operators to carry out their own analysis (Weeden 2012). While this level of secrecy may have been reasonable and practical when the United States was the sole source of such data and analysis, the national security imperatives are much less strong when other nations and commercial entities offer independent data sources and capabilities, and the need for high-quality data was much greater now that commercial space activity had grown (Borowitz 2018). Weeden and others argued that the most productive way forward was to remove the SSA data sharing task from the U.S. military and give it to a civil agency (Weeden 2016).

Both the DoD and the Department of Transportation (whose Office of Commercial Space Transportation was originally tapped as the potential owner of the civil system) were supportive of the idea. DoD officials argued that this division of labor would allow the military to focus on national security imperatives in space, including defending space assets and attributing bad actors. The Department of Transportation indicated that it was prepared to take on the new responsibility (Foust 2016). Despite the support by both agencies, Congress did not provide the funding and authority necessary to enable this transition.

The Trump administration, which began in 2017, revisited this debate. In 2018, the administration released Space Policy Directive 3 (SPD-3), the National Space Traffic Management Policy. This document discussed the importance of SSA for both safety and security. It called for continued U.S. provision of SSA data, free of charge, to users around the world and identified the Department of Commerce as the home for the civil SSA mission (United States 2018). While the DoD remained supportive and the Department of Commerce was enthusiastic about taking on this role, the transition of authority and funding remained stalled, largely due to debates within Congress about whether the Department of Transportation or Commerce was the most appropriate choice (Young 2021).

At a meeting of the National Space Council during which SPD-3 was to be discussed, President Trump surprised attendees by calling for the establishment of a Space Force as a sixth branch of the Armed Forces. Although the concept was new to the public, the creation of a Space Corps had been promoted by members of the House Armed Services Committee in 2017, and the DoD had already been tasked to develop a report on the subject (Becker 2018). In August 2019, President Trump re-established USSPACECOM as the eleventh Unified Combatant Command, an organization that would bring together troops to conduct operations in, from, and through space (DOD 2019). Not long after, in December 2019, Congress established the Space Force as an independent service within the Air Force (similar to the organization of the Marines within the Navy; Kennedy 2019).

Amidst these organizational changes, Air Force Space Command introduced a change in terminology: from SSA to SDA. This change was meant to emphasize the fact that the U.S. military views space as a domain of warfare, just like land, air, and sea. The term SSA had become associated with detecting and tracking objects in space to enable space safety. Military leaders explained that the term SDA was meant to help "shift our focus beyond a Space Situational Awareness mindset of a benign environment" (Erwin 2019). SDA would include existing SSA-type observations as well as intelligence and environmental monitoring to enable space battle management (Erwin 2019).

Aggressive behavior in space continued throughout this period. In March 2019, shortly before the re-organization of U.S. military space entities, India carried out a kinetic anti-satellite test, shooting down its own satellite in low Earth orbit. The U.S. response to this event was relatively muted, with a State Department official simply noting that the issue of space debris was an important concern (Staff 2019). Shortly after this change, U.S. officials called attention to the type of threatening behavior that SDA seeks to identify. Two Russian satellites had been "shadowing" a U.S. reconnaissance satellite (Hitchens 2020). One of the spacecraft later ejected a smaller spacecraft in a move interpreted by USSPACECOM as a weapons test. U.S. officials argued that this test showed that "threats to U.S. and Allied space systems are real, serious, and increasing," and highlighted the importance of U.S. Space Force and USSPACECOM (Hitchens 2020).

In June 2020, the DoD released the Defense Space Strategy, which reiterated that space is a distinct warfighting domain. The document directly highlighted the threats posed by China and Russia, demonstrated by their deployment of counterspace capabilities and doctrine regarding the use of counterspace capabilities. The document argues, "China and Russia each have weaponized space as a means to reduce U.S. and allied military effectiveness and challenge our freedom of operation in space" (DOD 2020: 1).

Despite these statements, the report largely emphasized the U.S. desire to maintain stability in space and repeatedly emphasized engagement with allies and the commercial sector. An example is the definition of space superiority given in the document, which states:

> DoD will establish, maintain, and preserve U.S. freedom of operations in the space domain. DoD will be prepared to protect and defend U.S. and, as directed, allied, partner, and commercial space capabilities and to deter and defeat adversary hostile use of space.
>
> (DOD 2020: 2)

In order to "shape the strategic environment," the document calls on the DoD to enhance stability and reduce the potential for miscalculations by partnering with the Department of State to work closely with allies and partners to develop common understandings of appropriate behavior in space (DOD 2020: 8).

The U.S. Space Force's publication "Spacepower," also released in June 2020, takes a similarly wholistic view to U.S. space security. The document begins with a quote from President Kennedy, rather than a military leader, and emphasizes the importance of space for exploration, science, and everyday life. It places blame on potential adversaries, whose "actions have significantly increased the likelihood of warfare in the space domain" (USSF 2020: iv). The United States, by contrast, "desires a peaceful, secure, stable, and accessible space domain" (USSF 2020: vi) and states that space systems must be protected "from those who would wish to harm them" (USSF 2020: v).

The document distinguishes between national spacepower, which encompasses all of a nation's efforts to use space for prosperity and security, and military space-power, which exists to preserve that prosperity and security. To achieve this, mili-tary space forces "deter aggression and apply lethal and nonlethal force in, from, and to space" (USSF 2020: 21). The document asserts that military spacepower has the potential to be the difference between victory and defeat in war, providing a strategic imperative for the existence of the U.S. Space Force (USSF 2020: 26).

The document identifies SDA as one of five core competencies for the U.S. Space Force. It provides a broad definition, stating, "Space Domain Awareness (SDA) encompasses the effective identification, characterization, and understand-ing of any factor associated with the space domain that could affect space opera-tions and thereby impacting the security, safety, economy, or environment of our Nation" (USSF 2020: 34). The document calls for leveraging allies, civil and industry partners. It emphasizes that SDA encompasses not only physical loca-tion but also the understanding necessary to detect deceit and determine adver-sary intentions (USSF 2020: 39). The emphasis on protecting and defending space assets, rather than a more aggressive offensive focus, was further reflected in the December 2020 announcement that members of the Space Force would be referred to as "Guardians" (Schwartz 2020).

While these founding documents had been released during the Trump adminis-tration, the Biden Administration largely continued this rhetorical approach in its first official space policy document, the United States Space Priorities Framework. The document recognizes the importance of space for the American way of life and acknowledges risks posed by counterspace capabilities of potential adversaries. It argues that conflict is not inevitable and that the United States will enhance the resilience of its space architecture, as well as the ability to detect and attribute hos-tile acts in space – a capability enabled by SDA. It calls not only for coordination with allies, partners, and commercial entities but also for diplomatic engagement with "strategic competitors" to enhance stability in outer space. The document calls for expanded SSA data sharing with all space operators facilitated by a civil agency (United States 2021).

In July 2021, for the first time in its history, the DoD put out an unclassified statement identifying a set of norms of behavior for military activities in outer space. While relatively vague, these guidelines reinforced existing tenants of the Outer Space Treaty and demonstrated the willingness of the U.S. military to engage more concretely on the issue of responsible behavior in space (Hitchens 2021). Russia demonstrated remaining challenges in this area with its kinetic anti-satellite test in November 2021, which created more than 1,500 new pieces of debris (Raju 2021).

While DoD officials still support the transition of the SSA data sharing respon-sibilities to a civil agency, they have continued to engage with international and commercial partners. Officials increased transparency, adding tracking data on some previously classified U.S. space objects, although others remain obfuscated (Hitchens 2019). By 2021, the USSPACECOM had signed more than 100 SSA Data Sharing Agreements with commercial entities, foreign nations, and inter-governmental organizations (USSPACECOM 2021). USSPACECOM maintains a multinational space collaboration cell with members from France, Germany, and the United Kingdom and plans to add additional members. It is also growing its commercial integration cell, which includes a number of commercial space entities that provide data to help improve the military's SSA (Erwin 2021).

Discussion

The detailed case study above shows that the U.S. military mission of tracking objects in space has a long history with numerous changes in terminology, program organization, and focus. However, as a whole – and perhaps not surprisingly – the program has almost always maintained a primary focus on the national security aspects of monitoring space activity. From the beginning of the space age to the early 2000s, the military spoke about monitoring objects in space almost exclusively with regard to the importance of identifying threats and enabling a response to those threats.

The engagement of the U.S. military with the space safety mission is only a relatively recent phenomenon, coinciding with the creation of the CFE pilot program in 2004. Even this transition was driven by a desire to retain control over data seen to have significant national security value. However, as global civil and commercial activity in space grew, this space safety mission required increased effort and gained significant attention domestically and internationally. These realities, combined with the Obama Administration's focus on international engagement and cooperation, created an environment in which the military spoke frequently of its contributions to space safety and sustainability.

However, as the scope and complexity of the space safety mission continued to grow in parallel with the growth of global space activity, its maintenance by the U.S. military became untenable. Providing civil and commercial users with SSA data could not be accomplished as a by-product of the military mission. Unlike the provision of a signal from a system like GPS, provision of SSA data requires that analysis capabilities scale with the number of space objects and data users. Also unlike GPS, provision of SSA capabilities requires continuous interaction with space operators to provide further analysis and input when conjunctions are possible.

Even before the Obama administration ended, the cracks in this system were showing. While its rhetoric reflected an acceptance of its space safety and sustainability mission, it struggled to fully meet the demand of civil and commercial space operators. The U.S. military still tended to emphasize secrecy, and threatening actions undertaken by potential adversaries prompted a desire to return to a focus on threat detection and enabling counterspace operations. DoD officials were eager to offload their civil and commercial-focused duties.

When USSPACECOM and U.S. Space Force were created, and the change in terminology from SSA to SDA occurred, this largely represented a continuation of a trend that had begun in the mid-2010s, at the end of the Obama administration. The move away from the space safety mission was not an unprecedented or aggressive move toward militarization, but rather a return to the U.S. military's long-standing focus on the national security role of SSA.

While the United States has been vocal about threats posed by potential adversaries, these have been primarily incidents of highlighting specific threatening actions, rather than theoretical threats that could emerge. Furthermore, this transition has not been accompanied by an emphasis on developing new offensive space weapons, as realists might expect. Early U.S. Space Force documents, developed under the Trump Administration, reflected a much more wholistic view of the value of space – emphasizing both the economic and security importance of space operations and embracing the importance of promoting space stability and sustainability. The U. S. military seems to be embracing traditional liberal approaches, including international cooperation and the development of norms of behavior.

The space environment of the 2020s is fundamentally different from that of the early 2000s, and while the U.S. military may be returning to a narrower national security focus, as symbolized by the adoption of the term "space domain awareness," it is doing so with an awareness that it must take into account the vast civil and commercial value of space for users around the world. Furthermore, the United States as a whole is embracing the space safety mission, not abdicating its role in this area. While domestic politics and bureaucratic inefficiencies may be delaying its implementation, the United States has committed to continue providing SSA data and services to users around the world free of charge. The implementation of this program within a civil agency will increase transparency, facilitate greater cooperation with international and commercial partners, and ultimately enable a higher level of service, improving prospects for space safety and sustainability.

Conclusion

This chapter sought to examine whether the recent U.S. military transition from using the term "space situational awareness" to "space domain awareness" symbolized a larger transition toward the militarization of space. By examining the history of U.S. military activity related to SSA throughout the space age, it became clear that the recent focus on the military aspects of space (including SDA) is a return to the long-standing approach the U.S. military has taken toward this issue, rather than a truly new development.

Even so, this narrower focus on national security space is understood by the U.S. military to exist within the context of the current space environment – one in which space assets are essential to everyday life around the world, and foreign and commercial use of space continues to expand rapidly. The U.S. Space Force, while recognizing the need to respond to counterspace threats, also acknowledges that space stability and sustainability are key national goals that it must help to protect.

The new willingness of the U.S. military to engage in international discussions regarding norms of behavior and the emphasis of the Biden administration on international engagement – even with "strategic competitors" – are trends to be encouraged. Threats in space are real, with rendezvous and proximity operations and anti-satellite tests increasing in recent years. While the United States will continue to develop military means to defeat these threats, the country and all other space actors should prioritize cooperation and engagement to discourage further militarization or weaponization of space, ensuring the domain remains stable and usable for all space actors.

Note

1 Despite a call for elevating the role of space activities and management within the DoD, U.S. Space Command was dis-established in October 2002 as a part of a post-September eleventh restructuring effort.

Bibliography

Aldridge Jr, E. C. (1987). "The Myths of Militarization of Space." *International Security* 11(4): 151–156.

Becker, R. (2018). *Trump Directs DOD to Establish a Space Force in a Surprise Announcement Today*. The Verge.

Borowitz, M. (2018). *Strategic Implications of the Proliferation of Space Situational Awareness Technology and Information: Lessons Learned from the Remote Sensing Sector*. Space Policy.

Brinton, T. (2008). *Shelton Urges Satellite Community to Share Data*. Space News.

Chow, T. (2011). *Space Situational Awareness Sharing Program: An SWF Issue Brief*. Secure World Foundation, TR, Washington, DC.

Congress, U. S. (2003). "National Defense Authorization Act (NDAA) for Fiscal Year 2004. U. S. Congress." *Public Law* 108–136.

DeFrieze, D. C. (2014). "Defining and Regulating the Weaponization of Space." *Joint Force Quarterly* 74: 110–115.

De Selding, P. (2008). *ESA Approves Space Situational Awareness Program*. Space News.

DOD. (2001). *Quadrennial Space Review Report*. United States Department of Defense.

DOD. (2010). *Quadrennial Defense Review Report*. United States Department of Defense.

DOD. (2019). *Department of Defense Establishes U.S. Space Command*. United States Department of Defense.

DOD. (2020). *Defense Space Strategy Summary*. United States Department of Defense.

Erwin, S. (2019). *Air Force: SSA Is No More; It's 'Space Domain Awareness'*. Space News.

Erwin, S. (2021). *Space Command to Expand Network of Allies that Help Monitor Orbital Traffic*. Space News.

Everett, T. (2007). *OpEd: America's Vulnerability in Space*. Space News.

Ferster, W. (2012). *JSPOC Conjunction Alerts Could Be Improved, Group Says*. Space News.

Foust, J. (2016). *FAA and Pentagon Foresee Gradual Transition of Space Traffic Management Activities*. Space News.

GAO. (1986). *Case Study of the Air Force Space Based Space Surveillance System*. United States General Accounting Office

GAO. (1987). *Strategic Defense Initiative Program: Status of Space Surveillance and Tracking System*. United States General Accounting Office

GAO. (1990). *Space Debris a Potential Threat to Space Station and Shuttle*. United States General Accounting Office.

GAO. (1997). *Space Surveillance: DOD and NASA Need Consolidated Requirements and a Coordinated Plan*. United States General Accounting Office.

GAO. (2002a). *Space Surveillance Network: Appropriate Controls Needed Over Data Access*. United States General Accounting Office.

GAO. (2002b). *Space Surveillance Network: New Way Proposed To Support Commercial and Foreign Entities*. United States General Accounting Office.

GAO. (2011). *Space Acquisitions: Development and Oversight Challenges in Delivering Improved Space Situational Awareness Capabilities*. United States Government Accountability Office.

Grondin, D. (2009). *The (Power) Politics of Space: The US Astropolitical Discourse on Global Dominance in the War on Terror*. Securing Outer Space, Routledge: 116–135.

Gruss, M. (2015). *Russian Satellite Maneuvers, Silence Worry Intelsat*. Space News.

Hansel, M. (2010). "The USA and Arms Control in Space: An IR Analysis." *Space Policy* 26(2): 91–98.

Hitchens, T. (2019a). *Intel Community's Secrecy Culture Frustrates DoD Sat Safety Effort*. Breaking Defense.

Hitchens, T. (2019b). *The Stellar Dance: US, Russia Satellites Make Potentially Risky Close Approaches*. Breaking Defense.

Hitchens, T. (2020a). *Lamborn, Horn Pledge Bipartisan Support For Space*. Breaking Defense.

Hitchens, T. (2020b). *Russian Sat Spits Out High-Speed Object In Likely ASAT Test*. Breaking Defense.

Hitchens, T. (2021). *Exclusive: In A First, SecDef Pledges DoD To Space Norms*. Breaking Defense.

Johnson, N. L. (2021). "Operation Burnt Frost: A View From Inside." *Space Policy* 56: 101411.

Johnson-Freese, J. and D. Burbach. (2019). "The Outer Space Treaty and the Weaponization of Space." *Bulletin of the Atomic Scientists* 75(4): 137–141.

Kennedy, M. (2019). *Trump Created The Space Force. Here's What It Will Actually Do*. NPR.

Krepon, M. and M. Katz-Hyman. (2007). *OpEd: An Arms Race in Space Isn't the Problem*. Space News.

McGlade, D. (2007). *OpEd: Preserving the Space Environment*. Space News.

Moltz, J. C. (2020). *The Politics of Space Security. The Politics of Space Security*. Stanford University Press.

NSC. (1955). *U.S. Scientific Satellite Program*.

Peoples, C. (2010). "The Growing 'Securitization' of Outer Space." *Space Policy* 26(4): 205–208.

Peter, L. (2015). *Russia Shrugs off US Anxiety Over Military Satellite*. BBC.

Posen, B. R. (2003). "Command of the Commons: The Military Foundation of US Hegemony." *International Security* 28(1): 5–46.

Raju, N. (2021). *Russia's Anti-Satellite Test Should Lead to a Multilateral Ban*. Stockholm International Peace Research Institute.

Raymond, J. (2007). *Operations Group Blazes New Trail During Operation Burnt Frost*. Peterson Air Force Base News.

Roberts, T. (2021a). *Unusual Behavior in GEO: Luch (Olymp-K)*. Center for Strategic and International Studies (CSIS).

Roberts, T. (2021b). *Unusual Behavior in GEO: SJ-17*. Center for Strategic and International Studies (CSIS).

Rosas, A. (1983). "The Militarization of Space and International Law." *Journal of Peace Research* 20(4): 357–364.

Rumsfeld, D., D. Andrews, R. Davis, H. Estes, R. Fogleman, J. Garner, W. Graham, C. Horner, D. Jeremiah and T. Moorman. (2001). *Report of the Commission to Assess United States National Security Space Management and Organization*. Government Printing Office, Washington, DC.

Schwartz, M. (2020). *Space Force Members Will Be Called 'Guardians'*. NPR.

SDA. (2022). "Welcome to the Space Data Association." Retrieved 5 February 2022, from https://www.space-data.org/sda/.

Shalal-Esa, A. (2009). *Pentagon May Reach Satellite Analysis Goal Early*. Reuters.

Singer, J. (2007). *Chilton: Progress Is Being Made On Space Situational Awareness*. Space News.

Singer, J. (2009). *Proposed SSA Improvements Include Non-Air Force Data*. Space News.

Space News Editor. (2007). *Editorial: For Space Surveillance, It's a Brave New World*. Space News.

Space News Staff. (2011). *Stratcom Could Negotiate Data Sharing Agreements*.

Staff. (2019). *US Says "Took Note" Of India's Statements On Debris Created By ASAT Test*. NDTV.

Sturdevant, R. W. (2008). "From Satellite Tracking to Space Situational Awareness: The USAF and Space Surveillance, 1957–2007." *Air Power History* 55(4): 4–23.

UN. (2002). *Conference on Disarmament Concludes Second Part of 2002 Session*. United Nations Conference on Disarmament.

United States. (2006). *U.S. National Space Policy*. United States Executive Office of the President.

United States. (2010). *U.S. National Space Policy of the United States of America*. United States Executive Office of the President.

United States. (January 2011). *National Security Space Strategy Unclassified Summary*. U.S. Department of Defense and U.S. Office of the Director of National Intelligence.

United States. (2018). *Space Policy Directive-3 (SPD-3)*. National Space Traffic Management Policy. Executive Office of the President.

United States. (2021). *United States Space Priorities Framework*.USAF (2004). *Counterspace Operations*. United States Air Force.

USAF. (2022). "Air Force Space Command History." Retrieved October 4, 2021, from https://www.afspc.af.mil/About-Us/AFSPC-History/.

USSF. (2020). *Space Capstone Publication: Spacepower: Doctrine for Space Forces*. United States Space Force.

USSF. (2022). "United States Space Force Mission." Retrieved from https://www.spaceforce.mil/About-Us/About-Space-Force/Mission/.

USSPACECOM. (1997). *Vision for 2020*. United States Space Command.

USSPACECOM. (1998). *Long Range Plan: Implementing USSPACECOM Vision for 2020*. United States Space Command.

USSPACECOM. (2021). *USSPACECOM Signs 100th Commercial Agreement to Share Space Data, Service*. United States Space Command Public Affairs Office.

Weeden, B. (2012). *Space Situational Awareness Is Bigger than U.S. Military*. Space News.

Weeden, B. (2016). *Time for the U.S. Military to Let Go of the Civil Space Situational Awareness Mission*. Space News.

Weeden, B. and V. Samson. (2021). *Appendix 01 – Historical Anti-Satellite Tests in Space by Country. Global Counterspace Capabilities: An Open Source Assessment*. Secure World Foundation.

Young, M. (2021). *Bad Idea: Stalling Funds for Commerce's Takeover of the Space Situational Awareness Catalog*. Center for Strategic and International Studies.

10   China and India as Rising Powers and the Militarisation of Space*

Dimitrios Stroikos

Introduction

This chapter offers a discussion of China and India as space powers with a specific focus on their growing interest in the use of space for military purposes. China and India are the two most consequential non-Western space powers, and both have made significant strides in developing their space capabilities over the past few decades, with Beijing in particular making important progress in recent years. Both countries have launched satellites for a range of dual-use applications and have demonstrated their ability to conduct destructive anti-satellite (ASAT) tests, with China conducting a successful test in 2007 and India in 2019.

This chapter starts with offering a background and overview of China's space programme that highlights the key achievements of its civilian programme before moving on to consider the ways in which China uses space for military purposes. Next, it provides background on India's civilian and military programme. The third section applies Scott Sagan's analytical framework on nuclear proliferation to China and India's military space activities by focusing on their ASAT tests. In his influential framework on nuclear proliferation, Sagan (1996–1997) puts forward three models that help explain why states develop nuclear weapons: the 'security model', the 'domestic politics model', and the 'norms model'. The security model is premised on the idea that a state's decision to acquire nuclear weapons is the result of national security considerations along the lines of realist thinking. The domestic politics model maintains that nuclear weapons serve not only military but also political purposes, which means that they can be used to promote narrow-minded domestic and bureaucratic interests. According to the norms model, states decide to either build nuclear weapons or abstain from their possession because of the importance of nuclear weapons as a 'normative symbol of a state's modernity and identity' (1996–1997: 55). In doing so, the author identifies three alternative reasons why states pursue nuclear weapons. Similarly, this chapter argues that when applied to China and India's ASATs, such a framework has the advantage of accounting for three different reasons why states build ASATs and the interplay between them by discussing key insights that this approach can bring into the study of China and India's ASATs.

DOI: 10.4324/9781003230670-14

China in Space: Background and Brief Overview

China's space programme has made significant strides in recent years, consolidating its role as a major space player in the current global space order. Although the story of China's space programme has been frequently told, a few salient points are worth briefly re-capping for the purposes of this discussion.[1] First, it is important to note that since its inception in the mid-1950s, China's space effort has been driven by the same priorities and rationales that underpinned the space programmes of the United States and the Soviet Union during the Cold War, including military, domestic political, and prestige considerations. After all, it needs to be remembered that the Chinese leadership had embarked upon the initial phases of a human spaceflight programme in the early 1970s, but by and by Mao decided to cancel it (Handberg and Li, 2006). Still, this is testimony to China's long-term ambitions in space.

Second, notwithstanding the political upheavals that defined the Great Leap Forward and the Cultural Revolution, some milestones were achieved under Mao, such as the successful launch of China's first satellite, Dong Fang Hong 1, in 1970, making it the fifth country to successfully launch a satellite into orbit. Given the limited resources and the social and political tumult of this period in China, however, the trajectory of the Chinese space programme might have been different were it not for the support from Zhou Enlai and others who strived to keep the strategic weapons programme on track by offering protection to the principal scientists and engineers working on the space programme (Chen, 1991).

Third, the launch of China's first satellite was part of the so-called 'Two Bombs, One Satellite' project that refers to the atomic and hydrogen bombs and the first satellite. The project was initiated in the late 1950s and early 1960s and is associated with a techno-nationalist approach to technological advancement (Cheung, 2009; Feigenbaum, 2003).[2] Although the focus of the Chinese space programme shifted to economic development under Deng Xiaoping, an attempt was made to revitalise the spirit and approach of the 'Two Bombs, One Satellite' project when in 1986 the '863 High-Technology Research and Development Plan' was introduced, which identified main areas of strategic significance, such as space.[3] It was against this backdrop that key space projects assumed renewed interest, including human spaceflight (Kulacki and Lewis, 2009: 21–24; Li, Ma, and Li, 2022). It is also useful to note that the 'Two Bombs, One Satellite' remains an important source of national pride, an influential top-down model of technological innovation, and a key part of the country's history and identity today, illustrated by the frequent references made to it by Chinese leaders. For example, in May 2022, Chinese President Xi encouraged the new generation of professionals in the Chinese aerospace sector 'to carry forward the spirit of the "Two Bombs, One Satellite"' in order to realise 'self-reliance in aerospace science and technology through innovation' (Xinhua, 2022). Reflecting the importance that Xi attaches to space, the Chinese President has also stressed how the country's space effort is part of the 'Chinese Dream' of rejuvenating the nation.[4] In his words, 'The space dream is part of the dream to make China stronger' (Xinhua, 2017).

Nevertheless, in recent years, the programme has seen a number of successes, such as the launch of Shenzhou 5 in October 2003, which sent China's first astronaut, Yang Liwei, into space and made China the third country to independently send a human into space. The Shenzhou project, China's human spaceflight programme, which has carried out a series of human spaceflight missions since then, has been part of a broader plan to establish a more robust presence in space through a space station and its lunar exploration programme. More specifically, in September 2011, China launched Tiangong 1 (Heavenly Palace), its first space laboratory module, followed by the launch of Tiangong 2 in 2016. These modules were designed as experimental test beds for the construction of the Tiangong space station, its larger orbital space station, which was completed in 2022.[5]

Another important aspect of China's space programme is lunar exploration. The Chinese lunar programme, known as the Chang'e programme, named after the mythical Chinese goddess of the Moon, has carried out a number of successful missions since Chang'e 1, China's first lunar mission that lasted from 2007 to 2009. In January 2019, China achieved a remarkable feat when the Chang'e 4 spacecraft landed on the far side of the Moon, becoming the first spacecraft to accomplish this, and allowing China to explore this previously uncharted area of the Moon's surface. Following that, in December 2020, China successfully completed the Chang'e 5 lunar sample return mission, the first mission in over 40 years to collect lunar samples and bring them back to Earth for scientific analysis. Likewise, in July 2020, China launched its first Mars mission, Tianwen-1, which entered orbit around Mars in February 2021.

Although China's achievements in space exploration have attracted much attention, these have been accompanied by a focus on satellite and rocket development. China has launched and operates a series of communication satellites for various applications and different types of remote-sensing satellites for earth observation, meteorology, and disaster and environmental monitoring. More recently, a notable achievement was the launch of the last satellite of the BeiDou Navigation Satellite System in June 2020, which signified the completion of its own equivalent of the GPS system with global coverage. To support its vigorous and expanding space activities, China has also been developing new variants of launch vehicles, including the Long March rocket family which can carry payloads of different sizes and launch from different types of launch sites. Meanwhile, Beijing has also taken steps to commercialise its space activities with the increasing involvement of private companies in the China space sector (Chandrashekar, 2022; Harvey, 2019; Pollpeter et al., 2020).

To these aspects can be added China's emphasis on international cooperation that centres on the promotion of bilateral ties with a number of countries in Asia, Africa, Latin America, and Europe, and its growing participation in multilateral institutions.[6] Within the scope of international cooperation and diplomacy, China has also tried to play a leadership role at the regional level, typified by the establishment of the Asia-Pacific Space Cooperation Organization (APSCO) in 2008, an intergovernmental organisation aimed at promoting multilateral cooperation on space science and technology between its member states, with a focus on providing training, data sharing, and capacity-building programmes in various areas of space technology and applications. Located in Beijing, APSCO's membership includes eight countries: China, Bangladesh, Iran, Mongolia, Pakistan, Peru, Thailand, and Turkey.[7] China has also incorporated space into the Belt and Road Initiative (BRI) with the implementation of the 'Space Information Corridor' or 'Space Silk Road', and it has set up an emergency response mechanism that aims to share satellite-based data with BRI nations to assist in weather monitoring and disaster management (Jones, 2018).

Further illustrating its leadership ambitions, China together with the United Nations Office for Outer Space Affairs (UNOOSA) has established the 'United Nations/China Cooperation on the Utilization of the China Space Station' programme within the framework of the UNOOSA's Access to Space for All Initiative. The aim is to provide opportunities for scientists and researchers from around the world to conduct experiments and research on China's space station. Such developments underscore how Beijing recognises the need for assuming great power responsibilities in the space community of states through the provision of public goods that can confer it the status of great power in space (Stroikos, 2022a: 343–346).

China and the Military Uses of Space

China's military has played a crucial role in the development of China's space endeavour, overseeing both its military and civilian activities from the outset (Wu, 2022: 17). However, it is clear that space has assumed a more important place in China's broader military modernisation effort as a consequence of the country's expanding interests across the globe and in new domains of strategic interaction. As a result, China's global reach requires a national military force with the capacity to operate beyond its immediate periphery, marking a significant shift from the People's Liberation Army (PLA) traditional orientation towards maintaining internal security and protecting continental China to supporting missions at long distances (Wortzel, 2013).

Encompassing this shift of orientation has been an appreciation of the potential contribution of space as a force multiplier in modern warfare since the early 1990s. Indeed, the 1991 Gulf War, often labelled as the 'first space war' by experts, prompted China to fundamentally re-evaluate its military concepts, doctrines, and training, while also recognising the essential role that the effective use of space assets plays in modern warfare (Cheng, 2012: 57–58; Liao, 2005: 206, 208). This was evident in the introduction of the new doctrine of 'local wars under modern high-tech conditions', which was accompanied by a growing interest in the research and development of state-of-the-art technologies and weaponry, including 'building satellites, early warning and command systems, and advanced communication relay stations' (Shambaugh, 2004: 70).[8]

Other external influences in tandem with the so-called 'Revolution in Military Affairs' (RMA) have had the effect of further evolving Chinese strategic thinking that changed from 'limited war under high-technology conditions' to 'local wars under modern informationalized conditions' in 2002 (Finkelstein, 2007: 104). This was followed by the announcement of the PLA's 'new historic missions' under President Hu Jintao in 2004, which, among other things, recognised for the first time that China's national interests were extended to outer space and the electromagnetic sphere (Cheng, 2012: 61).

Significantly, the strategic value attached to space in the context of China's military modernisation has increased under Chinese President Xi Jinping, who is also the chairman of the PLA Central Military Commission (CMC). In April 2014, during a visit to the PLA's air force headquarters in Beijing, Xi called for accelerating the integration of air and space capabilities (Zhao, 2014). Subsequently, China's 2015 defence white paper described outer space as a 'commanding height in international strategic competition' (State Council of the People's Republic of China, 2015). Similarly, the 2019 defence white paper refers to 'China's security interests in outer space' as one of its 'national defense aims' and that '[o]uter space is a critical domain in international strategic competition' (State Council of the People's Republic of China, 2019).

Meanwhile, under Xi, there has been a major restructuring of the Chinese military forces that led to the creation of the People's Liberation Army Strategic Support Force (PLASSF) with the goal of improving the strategic-level coordination and efficiency of space, cyber, and electromagnetic operations in a more integrated manner.[9] Alongside this, since 2016, under Xi an attempt has been made to revamp civil-military integration, as part of the policy of military-civil fusion (MCF), with the aim of integrating the development and use of civilian technologies into military applications in order to improve China's national security and military capabilities. In this way, this policy entails leveraging the technological advancements made in the civilian and commercial sectors to enhance China's military modernisation and competitiveness by facilitating effective coordination between the government, military, and civilian sectors to attain technological breakthroughs and innovation. To this end, the Chinese government elevated civil-military fusion to a national-level strategy and formed a new specialised central commission, the Central Commission for Integrated Military and Civilian Development, in 2017, to oversee its implementation.[10] It is too early to assess the impact of this effort but it is clear that within the scope of MCF, emphasis has been placed on the space sector as a way to enhance coordination between the military and civilian sectors, facilitate resource and data sharing, and encourage participation from non-governmental entities. For some observers, this move could be a significant step towards the reconstruction of the Chinese space industry (Wu, 2022: 18).[11]

One of the principal manifestations of how China recognises the critical place of space for the PLA has been the emphasis on the use of space-based systems in support of Command, Control, Communications, Computers, Intelligence, Surveillance, and Reconnaissance (C4ISR) as part of networked integrated C4ISR architecture that will enable Chinese military forces to operate effectively in a variety of environments and situations. Apart from the deployment of dedicated military satellites, this has involved the development of a network of dual-use satellites with a variety of military functions that provide support for C4ISR, including the utilisation of communications satellites as well as a variety of high-resolution remote-sensing satellites, such as the *Yaogan, Haiyang, Huanjing,* and *Gaofen* series, Tracking and Data Relay Satellites (TDRS), and Beidou global positioning timing and navigation satellites.[12]

Furthermore, it is believed that a key component of how China prioritises the military usage of space is the development of counterspace capabilities. In particular, there appears to be evidence to support that China has been working on various technologies that have the ability to provide both destructive and non-destructive means of interfering with space activities, such as direct-ascent, co-orbital, electronic warfare, and directed energy weapons capabilities.[13] This was evident in January 2007, when the PLA carried out a direct-ascent ASAT that destroyed one of its own ageing weather satellites, FengYun-1C (FY-1C), making the country the third in the world to conduct this type of test. The test generated a large amount of space debris and was widely condemned by the international community. It was also inconsistent with China's long-standing position against space weaponisation. While possible explanations of this test are considered below, what should be mentioned here is that since after the 2007 ASAT test, China has also carried out a series of known or suspected non-destructive direct-ascent ASAT tests the most remarkable of which occurred in May 2014, when the launch of a missile is thought to have attained an altitude of about 30,000 km. This is seen as evidence of a new sophisticated ASAT weapons system that can effectively engage objects orbiting at medium or high altitudes above Earth (Weeden, 2014).

India in Space: Background and Brief Overview

Like China, there is now a growing body of literature that examines certain aspects of India's space programme and its history.[14] However, a few points are worth highlighting here for the purposes of this discussion. First, for some informed observers, the origins of India's space endeavour can be found in India's participation in the International Geophysical Year (IGY) of 1957–1958 (Kochhar, 2008; Reddy, 2008). It should be recalled that the IGY was a global scientific programme that engaged in investigating and understanding various aspects of the Earth's physical environment epitomising scientific internationalism and international cooperation, which led to the emergence of the Space Age with the launch of the first artificial satellite, Sputnik 1, in October 1957 (Stroikos, 2018). Notable Indian scientists, such as Vikram Sarabhai, played a significant role in the organisation and execution of crucial IGY undertakings. This initial participation served as a driving force for the development of India's space programme, under the leadership of Sarabhai (Kochhar, 2008). As a result, a nascent space programme was established in 1962 with the creation of the Indian National Committee for Space Research and the launching of the first Nike-Apache sounding rocket from the Thumba Equatorial Rocket Launching Station (TERLS) in 1963. This was followed by the formation of the Indian Space Research Organisation (ISRO) in 1969.

Second, one of the most notable features of India's space programme from the outset has been its priority to use space for socio-economic development. Unlike the historical trajectory of the space efforts of other space powers that benefited from military programmes, as Michael Sheehan (2007: 142–143) points out, 'the Indian space and missile development programmes are distinct enterprises, notwithstanding their common roots, and the technologies employed are far from interoperable'. In this respect, Indian scientists gave precedence to practical applications, including communications, meteorology, and remote sensing, to meet the social and developmental needs of the country (Sankar, 2007: 1–2). Therefore, underlying India's space effort was a development rationale and vision formulated by Sarabhai (1974). As Sarabhai (1968: 39) famously remarked:

> There are some who question the relevance of space activities in a developing nation. To us, there is no ambiguity of purpose. We do not have the fantasy of competing with the economically advanced nations in the exploration of the moon or the planets or manned space flight. But we are convinced that if we are to play a meaningful role nationally and in the community of nations, we must be second to none in the application of advanced technologies to the real problems of man and society, which we find in our country. The application of sophisticated technologies and methods of analysis to our problems is not to be confused with embarking on grandiose schemes whose primary impact is for show rather than for progress measured in hard economic and social terms.

Resting upon this rationale for the space programme and pivoted around meeting the socio-economic needs of a developing country, Sarabhai's vision was also entrenched in the idea of science and technology as 'as enablers of modernization, and particularly a belief in state management of resources directed towards "leapfrogging" stages of development into a state of modernity' (Siddiqi, 2015: 35). Such a vision was also aligned with Jawaharlal Nehru's enthusiasm for the role that science and technology can play as a progressive force in Indian society and was shared by other prominent scientists, such as Homi Bhabha, who supported the space programme (Siddiqi, 2015). In this context, the use of space was also viewed as a scientific pursuit of national significance that could allow India to become a technologically advanced country as well as symbol of 'nation-building' and a source of international prestige (Kochhar, 2008; Sheehan, 2007).

As a result, from the beginning, three key components provided the foundation for the development of India's space programme: communications and remote-sensing satellites; practical applications; and space transportation systems (Lele, 2021: 7). Regarding communications and remote-sensing satellites, the Indian National Satellite (INSAT) system became operational in 1983 and the first Indian Remote Sensing (IRS) Satellite, as part of the IRS satellite programme, was launched in 1988. In terms of applications, the focus has been on areas, such as meteorological observation and forecasting, disaster monitoring, tele-education and telemedicine, and the management of natural and earth resources (Reddy, 2008: 238). As per space transportation systems, India took steps to develop indigenous launch capabilities. Typifying this was the successful launch of the Satellite

Launch Vehicle (SLV)-3 in 1980, and the subsequent building of more sophisticated launch vehicles, such as the Augmented Satellite Launch Vehicles (ASLV), the Polar Satellite Launch Vehicle (PSLV), and the Geosynchronous Satellite Launch Vehicle (GSLV) (Lele, 2021: 23–40).

However, it is plain that India's ambitions as a great power in international society have also affected the overall trajectory of India's space programme. In reality, even though the development rationale remains a key element of India's space activities, there has been a notable shift in space exploration missions, exemplified by the launch of the country's first lunar mission called Chandrayaan-1 in 2008, which confirmed the presence of water molecules on the Moon's surface and marked a significant milestone in India's space programme. After the successful completion of the Chandrayaan-1 mission, ISRO was motivated to pursue further space exploration missions. The next natural progression was seen as a mission to Mars, building on a similar approach taken by other countries with established space programmes (Laxman, 2014: 70–72). Yet, as some observers have suggested, the timing of the decision and the limited scientific value of the Mars mission indicate that it was partly driven by political and strategic considerations in light of the unsuccessful Russian-Chinese Phobos-Grunt Mars mission in November 2011, as it would allow India to reach Mars ahead of China (Bagla and Menon, 2014: 47–49). Nevertheless, on November 5, 2013, India launched the Mars Orbiter Mission (MOM), also known as Mangalyaan, which resulted in India becoming the first country in Asia to successfully reach Mars. Remarkably, and emblematic of this reorientation towards space exploration, during his Independence Day speech in August 2018, Indian Prime Minister Narendra Modi, declared that India intends to launch its first human spaceflight mission by 2022, also known as the 'Gaganyaan' programme, but this has since been delayed and the country's maiden human spaceflight mission is now expected in 2024. In the meantime, ISRO is working on other future space missions, while significant steps have been taken to promote the commercialisation and privatisation of the Indian space sector.

India and Space Militarisation

Coupled with this reorientation towards space exploration, another key dimension of India's space programme has been the increasing interest in military uses of space. Nowhere has this been more apparent than in India's ASAT test, named 'Mission Shakti', which was carried out in March 2019, when a missile was launched from the Kalam Island missile complex that managed to intercept an Indian military satellite called Microsat-R. The satellite was in low orbit located about 300 km above the Earth's surface, built by the Defence Research and Development Organisation (DRDO) and launched by ISRO in January 2019 (Weeden and Samson, 2021: 5–3). Different explanations of the test are discussed in detail in the next section. But the point to make at this stage is that the militarisation of India's space programme has also been evident in the deployment of various satellites that have become crucial as a 'force multiplier' for Indian military capabilities. Therefore, since 2000, ISRO has put several remote-sensing satellites in orbit that have dual-purpose applications, with the ability to generate very high-resolution images suitable for military applications (Paracha, 2013: 164). Part of the reason for this can be attributed to the fact that the 1999 Kargil War revealed to the Indian military the insufficiency of India's remote-sensing infrastructure in offering efficient surveillance and timely identification of hostile incursions (Gopalaswamy, 2019: 80). More specifically, as Ajey Lele (2011: 384) notes, some of the India Remote Sensing (IRS) series satellites launched after 2000 are being considered to have dual-use utility. However, the military value is clearer when it comes to the high-resolution CARTOSAT (Cartographic satellite) series. In addition, the RISAT (Radar Imaging Satellite) series of remote-sensing satellites launched by India since 2009 uses synthetic aperture radar (SAR) to capture high-resolution images of the Earth's surface with dual-use applications (Paracha, 2013: 165).

India has also developed an independent regional navigation satellite system called NavIC (Navigation with Indian Constellation), also known as the Indian Regional Navigation Satellite System (IRNSS), which offers accurate positioning, navigation, and timing information to civilian and military users. The Standard Position Service (SPS) provides services to civilian users, and the Restricted Service (RS) is encrypted and available only to authorised military users.[15] Related to this, ISRO has also established jointly with the Airport Authority of India (AAI), the GPS Aided Geo Augmented Navigation (GAGAN) system, which has some dual-use utility (Gopalaswamy, 2019: 116).

A further dimension of India's interest in the military uses of space concerns the launch of dedicated military satellites. For example, in 2013, ISRO launched India's first dedicated military communications satellite GSAT-7, designed to improve the ability of the Indian Navy to carry out maritime operations by enhancing its surveillance capabilities. Likewise, in 2018, ISRO launched GSAT-7A, a military communication satellite, intended to augment the communication capabilities of the Indian Air Force (IAF). Following that, in 2019, India launched its first Electro-Magnetic Intelligence Satellite (EMISAT), a sophisticated electronic intelligence (ELINT) satellite developed jointly by ISRO and the Defence Research and Development Organisation(DRDO), to improve the ability of Indian Armed Forces to identify and intercept signals transmitted by hostile radar systems (Rajagopalan, 2020).

At the same time, encapsulating the need to integrate the use of space into military operations, in 2010 India formed the Integrated Space Cell under the command of the Integrated Defence Services (IDS) Headquarters of the Ministry of Defence as a nodal agency to coordinate and synergise the use of space assets and technologies for defence purposes, jointly operated by three services of the Indian Armed Forces (army, navy, and air force), the Department of Space, and ISRO (Weeden and Samson, 2021: 5–6). Further reflecting its recognition of the importance of space technology in national security, and soon after conducting the ASAT test in 2019, India created the Defence Space Agency (DSA), which is expected to play a key role in India's efforts to address space-based threats and boost its space power (Press Trust of India, 2019a). As well, the Indian government approved the establishment of the Defence Space Research Organisation

(DSRO), which will be involved in building space warfare capabilities and will offer technical and research support to the DSA (Raghuvanshi, 2019). Remarkably, in the same year, India announced plans to conduct its first simulated space warfare exercise, named 'IndSpaceEx' (Pandit, 2019). Complementing this greater focus on national security, ISRO recently inaugurated 'Project NETRA (Network for Space Objects, Tracking, and Analysis)', to track and analyse space objects. This system is intended to enhance India's space situational awareness capabilities and protect its space assets by providing early warning of potential threats (Madhumathi, 2019).

Beyond military space capabilities, to enhance its space power, especially vis-à-vis China, India has stepped up its military space ties with a number of strategic partners, such as the United States. For instance, in 2020, New Delhi entered into an agreement with Washington, the Basic Exchange and Cooperation Agreement on Geospatial Cooperation (BECA), which permits access to satellite data from the United States. This access to US satellite data will help India improve the precision of its weapons, including missiles and drones (Reuters, 2020). It is also worth noting that the Quadrilateral Security Dialogue (QSD), commonly referred to as the Quad, among Australia, India, Japan, and the United States, which is usually seen as a counterbalance to China, has recently added a space component with the formation of a new working group, the Quad Space Working Group, focusing on sharing satellite data. Moreover, Quad members have also announced a satellite-based maritime domain awareness initiative (Si-soo, 2022).

China, India, and ASAT Tests: A Three-Model Approach

The Security Model

Even though there has been a growing interest in the study of the politics of space from an International Relations theory perspective over the last decade or so, explanations of why states acquire space capabilities and counterspace weapons in particular, whether by practitioners or scholars, are still dominated by the neorealist or structural realist school of thought that is in accordance with security model.[16] Therefore, it is not surprising that structural realist perspectives have offered parsimonious and convincing explanations of China ASAT test. For example, according to Tellis (2007), China's decision to conduct the test was motivated by a strategic need to counterbalance the military space superiority of the United States by using asymmetric means, such as counterspace capabilities. In this way, as Tellis points out, it is strategically logical for Beijing to take advantage of Washington's increasing dependence on space-based military assets as a potential vulnerability or 'Achilles heel', especially in the event of a future conflict between China and the United States over Taiwan.

Pertaining to realist thinking is also the security dilemma that exists when one state's efforts to increase its security can lead to increased insecurity for others, triggering a spiral of action-reaction (Stroikos, 2022b). Indeed, it can be argued that space relations are more prone to security dilemmas due to the inherent dual-use nature of space technology (Johnson-Freese, 2007). Be that as it may, it is this action-reaction dynamic that leads some observers to suggest that a security dilemma in space has unfolded between the United States and China that can help to account for Beijing's decision to undertake an ASAT test in 2007. Within this action-reaction dynamic, China's interest in building counterspace capabilities can be seen as a response to US plans for space dominance and missile defence (Zhang, 2011). From a Chinese perspective, therefore, Chinese officials and experts came to believe that the weaponisation of space was bound to happen. They were also worried that US missile defence could potentially compromise their nuclear deterrent (Zhang, 2005; Zhang, 2011).

In a rather similar fashion, India's 2019 ASAT test can be understood from the lens of the security model. Immediately following the test, Indian Prime Minister, Narendra Modi, emphasised that the primary goal of the mission was to safeguard the security of the country, promote its economic growth, and advance India's technological advancement, and that it was not aimed at any specific country (Modi, 2019). And yet, as has been acknowledged by many observers, it was China's ASAT test in 2007 that prompted India's interest in counterspace capabilities, such as ASAT technology as well as shift in New Delhi's historical position towards the militarisation of space. To be sure, shortly after the Chinese test, Indian political leaders were adamant that New Delhi's position had not changed and that the country remained against the weaponisation of space. However, there were also several remarks made by high-ranking military and DRDO officials, including Vijay Kumar Saraswat, indicating India's intention to build ASAT capabilities using its missile defence system (Rajagopalan, 2011). Consequently, the Indian decision to undertake an ASAT test suggests that national security considerations vis-a-vis China have become an important factor. It also points to the existence of a security dilemma in space, even if not definitive, between Beijing and New Delhi (Lele, 2019).

The Domestic Politics Model

Alongside the security model, the domestic politics model with its focus on domestic political considerations and the role of bureaucratic interests is also useful in trying to explain China's 2007 ASAT test. Within this framework, for instance, Kulacki and Lewis (2008) offer an alternative explanation of China's decision to conduct an ASAT test in 2007 by arguing that it was less likely a reaction to US plans. Instead, looking at the decision-making process of the test, one possible explanation is that project managers may have been compelled to demonstrate to the Chinese leadership that the technology had reached a state of maturity and was now viable for use. As a result, these managers probably lobbied and succeeded in obtaining approval for the test. It is also reasonable to assume that the decision to aim at a satellite, as opposed to intercepting a missile, was made because the former was deemed to be a simpler undertaking. It is also important to remember that the Ministry of Foreign Affairs took a full 12 days to issue a statement acknowledging the test, which caused confusion among analysts as to whether Ministry of

Foreign Affairs officials were aware of the test. This delay raised concerns that the PLA may have carried out the test without obtaining the necessary approval from the Chinese security and foreign policy bureaucracy. Even though it is more likely that the Chinese leadership had approved the test, this situation suggests the presence of factionalism between the Ministry of Foreign Affairs and the PLA (Gill and Kleiber, 2007).

The domestic politics model is also useful in trying to understand India's ASAT test. In terms of bureaucratic politics and internal decision-making processes, it is not uncommon for influential technocratic bureaucracies like ISRO and DRDO to undertake initiatives without first securing political endorsement. These endeavours are intended to put pressure on the political leadership to eventually grant approval. The case of the Indian ASAT appears to be an example of this possibility, considering that officials associated with DRDO had been advocating for the development of ASAT capabilities for a while. They had indicated that India had all the necessary technological components for such a programme, which could be utilised when deemed appropriate by the political leadership (Gopalaswamy and Kampani, 2014: 55).

This perspective is also pertinent to the role of leaders. In this respect, we do know now that the DRDO had developed ASAT capability back in 2012. However, the prior administration did not authorise testing. This situation changed with the advent of the Modi government, which granted permission for the test to proceed (Tripathi, 2019). What is more, some believed that Modi's decision to proceed with Mission Shakti and publicly announce it during the 2019 general election campaign was motivated by domestic political considerations. For this reason, the opposition strongly condemned the test, alleging that Modi had attempted to exploit the mission for electoral gains (Press Trust India, 2019b). Finally, space achievements, including ASATs, can be an important source of national pride, and thus, they can contribute to bolstering the image of political leaders, such as Modi, or the legitimacy of regimes, such as the Chinese Communist Party.

The Norms Model

Under the norms model nuclear weapons can be considered as symbols of modernity and identity, and the same can be said about space weapons. A number of key insights can flow from a consideration of how this model can be related to the development and testing of ASATs. First, one of the most well-established insights that emerges from the constructivist literature is China and India's quest for great power status in international society.[17] In this regard, acquiring space technology can be seen as part of the 'recognition games' that states play in order to attain the status of great power in the space community of states. Consequently, obtaining ASAT capabilities can also be seen as a component of China and India's pursuit to establish themselves as great powers, highlighting the significance of the link between a country's sense of identity and its ability to develop space weapons in a hierarchical global space order. An important and revealing illustration of this possibility was Modi's televised address to the nation announcing the success of Mission Shakti. In his address, the Indian Prime Minister applauded the test as a big 'moment of pride for every Indian' that attested India's credentials as 'a global space power' 'using an indigenously developed' technology (Modi, 2019). Additionally, it is plausible to argue that New Delhi's interest and eventual decision to move on with the conduct of an ASAT test are reflective of India's bitter experience with the 1968 Nuclear Non-proliferation Treaty (NPT) (Samson 2010). From this perspective, to avoid a similar discrimination from happening this time in space with the creation of a potential hierarchical order between countries that possess ASAT weapons and those that do not, Mission Shakti cannot be understood without taking into account India's quest for prestige that comes with being part of a select group of nations possessing ASATs.

Second, history is also important in the context of discussing how national identity informs space behaviour. In the 19th century, scientific and technological progress took the form of a 'standard of civilisation' that differentiated European societies from non-European ones, premised on a hierarchical conception of science and technology that was embedded in a 'techno-scientific orientalist' discourse. This process had a significant impact on China and India in the sense that both countries continue to use visible manifestations of techno-scientific advancement, such as space projects, as indicators of power, status, and modernity. Typifying this is the continuing influence of a techno-nationalist ideology in China and India (Stroikos, 2020).

Third, great power aspirations mediated by identity are also pertinent to how norms of responsible behaviour can constrain or alter state preferences and actions. In a telling illustration of the potential impact of nascent norms of responsible behaviour and what the dynamic idea of a 'responsible great power' in space entails, after its 2007 ASAT test and the international condemnation it garnered, China conducted subsequent ASAT tests in a 'responsible' manner by avoiding the creation of space debris and labelling them as 'missile defense tests', influenced by the US Operation Burnt Frost in 2008. This suggests that Beijing was engaging in a process of social learning in that it has conformed to an emerging norm of environmental responsibility in space according to which the testing of ground-based kinetic ASAT capabilities by limiting or avoiding the generation of space debris is socially acceptable (Stroikos, 2022a: 345). Correspondingly, connoting how New Delhi conducted its ASAT test in a more 'responsible' way when compared to Beijing's 2007 test, Indian authorities were eager to emphasise that the Indian ASAT test was intentionally performed at a relatively low altitude. This would enable the space debris produced by the test to quickly re-enter the Earth's atmosphere, so it would not pose a significant danger to the space assets of other nations (Weeden and Samson, 2021: 5–4).

Conclusions

The aim of this chapter has been to provide an analysis of China and India as space powers with a focus on the growing militarisation of their space programmes. The chapter started with a discussion of China's civilian space activities as a necessary background, which involved highlighting the country's remarkable space feats in human spaceflight and lunar exploration, before considering the ways in which Beijing places an increasing emphasis on the use of space for military purposes. Then, it provided a brief overview of India's civilian space programme and its shift from the use of space for development to space exploration missions reflecting the ambitions of a great power in space. Like China, therefore, India's space programme has gone a long way since its inception. Furthermore, as this chapter has shown, this significant transformation of India's space programme has also been accompanied by a greater emphasis on military applications.

This chapter then moved on to consider the insights that can emerge from an application of Sagan's three models on nuclear proliferation to China and India's ASAT tests. Not surprisingly, perhaps, most analyses of China and India's ASAT tests have been in line with the security model. But although the security model is useful in explaining the drivers behind the Chinese and Indian ASATs, this chapter also shows the analytical utility of the domestic politics model by underscoring how domestic political considerations and bureaucratic interests have also been at play in both cases. Likewise, within the scope of the norms model, this chapter has illustrated the ways in which identity, history, and norms are important factors that shape and constrain the development and testing of ASATs. The key point is that rather than treating the three models as competing, their relationship should be seen as complementary by recognising how ASATs are dependent on the interaction of a complex array of factors and actors than a single model can accommodate. Such considerations assume renewed importance in an era of strategic uncertainty defined by great power competition and predominant assumptions about the inevitability of conflict in space. As a result, it is more important than ever to develop frameworks that help to capture the complex forces determining space policy outcomes, such as ASAT tests, with significant implications for the security and stability of the domain of space.

Acknowledgements

An earlier version of the chapter was presented at the 11th ESSCA Space Policy Workshop. I thank all the participants in the workshop for their helpful comments. I am also grateful to Thomas Hoerber and Iraklis Oikonomou, who read the manuscript and suggested many improvements.

Notes

* This chapter is a significantly revised and restructured version of a forthcoming work entitled 'Still Lost in Space? Understanding China and India's Anti-Satellite Tests Through an Eclectic Approach'.
1 For recent detailed accounts of China's space programme on which this section draws, see Aliberti (2015); Chandrashekar (2022); Harvey (2019); Pollpeter et al. (2020); and Wu (2022).
2 For a thoughtful analysis of Chinese techno-nationalism, see Hughes (2006).
3 On the importance of the '863 High Technology Plan', see Feigenbaum (2003).

4 For an insightful discussion of the China Dream and its meanings, see Callahan (2013). On Xi and his objective of making China great again, see Brown (2022).
5 For a recent account of China's space station, see Jiang and Zhao (2021).
6 For a recent discussion on China's space programme and international cooperation, see Wu (2022).
7 More information on APSCO can be found on the organisation's site: http://www.apsco.int.
8 For a detailed account of the effect of the Gulf War of 1991 and other external influences on the Chinese military, see Shambaugh (2004).
9 On the creation of the PLASSF and its importance, for example, see Costello and McReynolds (2018); and Pollpeter, Chase, and Heginbotham (2017).
10 For two insightful discussions of the MCF, see Bitzinger (2021) and Kania and Laskai (2021).
11 On space and MCF, also see: Nie (2020); and Wu and Long (2022).
12 On China's space capabilities and C4ISR, see Chandrashekar (2022); Pollpeter et al. (2020); and Wortzel (2013).
13 For two useful assessments of China's counterspace capabilities, see Weeden and Samson (2021); and Harrison et al. (2021).
14 On India's space programme, among others, see Aliberti (2018); Raj (2000); Rao and Radhakrishnan (2012) and Singh (2017).
15 More information on NavIC can be found on ISRO's site: https://www.isro.gov.in/SatelliteNavigationServices.html.
16 On International Relations theory and the study of space, including realism, see Stroikos (2022b). For a more extensive discussion of China and India's ASAT tests from an analytical eclectic approach that combines structural imperatives, domestic influences, and national identity, on which this section substantially relies, see Stroikos (forthcoming).
17 On China's quest for great power status, inter alia, see Deng (2008) and Suzuki (2008). For two recent works on India and status, see Basrur and Sullivan de Estrada (2017); and Schaffer and Schaffer (2016).

Bibliography

Aliberti, M. (2015). *When China Goes to the Moon...*. Cham: Springer.
Aliberti, M. (2018). *India in Space: Between Utility and Geopolitics*. Cham: Springer.
Bagla, P. and Menon. S. (2014). *Reaching for the Stars: India's Journey to Mars and Beyond*. New Delhi: Bloomsbury India.
Basrur, R. and Sullivan de Estrada, K. (2017). *Rising India: Status and Power*. London: Routledge
Bitzinger, R.A. (2021). China's Shift from Civil-Military Integration to Military-Civil Fusion. *Asia Policy*, 16(1), 5–24.
Brown, K. (2022). *Xi: A Study in Power*. London: Icon Books.
Callahan, W. A. (2013). *China Dreams: 20 Visions of the Future*. New York: Oxford University Press.
Chandrashekar, S. (2022). *China's Space Programme: From the Era of Mao Zedong to Xi Jinping*. Singapore: National Institute of Advanced Studies (NIAS) and Springer.
Chen, Y. (1991). China's Space Policy: A Historical Review. *Space Policy*, 7(2), 116–128.
Cheng, D. (2012). China's Military Role in Space. *Strategic Studies Quarterly*, 6(1), 55–77.
Cheung, T.M. (2009). *Fortifying China: The Struggle to Build a Modern Defense Economy*. Ithaca, NY: Cornell University Press.
Costello, J. and McReynolds, J. (2018, October 2). *China's Strategic Support Force: A Force for a New Era*. Washington, DC: National Defense University Press.

Deng, Y. (2008). *China's Struggle for Status: The Realignment of International Relations*. New York: Cambridge University Press

Feigenbaum, E. A. (2003). *China's Techno-Warriors: National Security and Strategic Competition from the Nuclear to the Information Age*. Stanford, CA: Stanford University Press.

Finkelstein, D. M. (2007). China's National Military Strategy: An Overview of the "Military Strategic Guidelines". In: R. Kamphausen and A. Scobell (Eds.), *Right Sizing the People's Liberation Army: Exploring the Contours of China's Military* (pp. 69–140). Carlisle: Strategic Studies Institute; US Army War College.

Gill, B. and Kleiber, M. (2007). China's Space Odyssey: What the Antisatellite Test Reveals about Decision-Making in Beijing. *Foreign Affairs*, 86(3), 2–6.

Gopalaswamy, B. (2019). *Final Frontier: India and Space Security*. Chennai: Westland Publications.

Gopalaswamy, B. and Kampani, G. (2014). India and Space Weaponization: Why Space Debris Trumps Kinetic Energy Antisatellite Weapons as the Principal Threat to Satellites. *India Review*, 13(1), 40–57.

Handberg, R. and Li, Z. (2006). *Chinese Space Policy: A Study in Domestic and International Politics*. Abingdon: Routledge.

Harrison. T., Johnson, K., Moye, J. and Young, M. (2021). *Space Threat Assessment 2021*. Washington, DC: CSIS. Available at: https://www.csis.org/analysis/space-threat-assessment-2021.

Harvey, B. (2019). *China in Space: The Great Leap Forward*. Second Edition. Chichester: Springer-Praxis.

Hughes, C. R. (2006). *Chinese Nationalism in the Global Era*. London: Routledge.

Jiang, S. and Zhao, Y. (2021). China's National Space Station: Opportunities, Challenges, and Solutions for International Cooperation. *Space Policy*, 57, 101439.

Johnson-Freese, J. (2007). *Space as a Strategic Asset*. New York: Columbia University Press.

Jones, A. (2018).Chinese Satellites to Provide Emergency Response for Belt and Road Countries. *FindChinaInfo*, 3 May. Available at: https://findchina.info/chinese-satellites-to-provide-emergency-response-for-belt-and-road-countries.

Kania, E. B. and Laskai, L. (2021). *Myths and Realities of China's Military-Civil Fusion Strategy*. Washington, DC: Center for a New American Security, January. Available at: https://www.cnas.org/publications/reports/myths-and-realities-of-chinas-military-civil-fusion-strategy.

Kochhar, R. (2008). Science as a Symbol of New Nationhood: India and the International Geophysical Year 1957–58. *Current Science*, 94(6), 813–816.

Kulacki, G. and Lewis, J.G. (2008). Understanding China's Antisatellite Test. *Nonproliferation Review*, 15(2), 335–347.

Kulacki, G. and Lewis, J.G. (2009). *A Place for One's Mat: China's Space Program, 1956–2003*. Cambridge, MA: American Academy of Arts and Sciences.

Laxman, S. (2014). *Indian Martian Odyssey: A Journey to the Red Planet*. New Delhi: Partridge India.

Lele, A. (2011). Indian Armed Forces and Space Technology. *India Review*, 10(4), 379–393.

Lele, A. (2019). Space Security Dilemma: India and China. *Astropolitics*, 17(1), 23–37.

Lele, A. (2021). *ISRO: Institutions that Shaped Modern India*. New Delhi: Rupa Publications India Pvt. Ltd.

Li, C., Ma, B. and Li, X. (2022). The Decision-Making Process of China's Human Spaceflight Program. *Space Policy*, 61, 101492.

Liao, S. (2005). Will China Become a Military Space Superpower? *Space Policy*, 21(3), 205–212.

Madhumathi, D.S. (2019). ISRO Initiates 'Project NETRA' to Safeguard Indian Space Assets from Debris and other Harm. *The Hindu*, 24 September. Available at: https://www.thehindu.com/sci-tech/science/isro-initiates-project-netra-to-safeguard-indian-space-assets-from-debris-and-other-harm/article29497795.ece.

Manoranjan Rao, P.V. and Radhakrishnan, P. (2012). *A Brief History of Rocketry in ISRO*. Hyderabad: Universities Press.

Modi, N. (2019). Speech by Prime Minister on "Mission Shakti", India's Anti-Satellite Missile test conducted on 27 March, 2019. Ministry of External Affairs, Government of India. Available at: https://www.mea.gov.in/Speeches-Statements.htm?dtl/31180.

Nie, M. (2020). Space Privatization in China's National Strategy of Military-Civilian Integration: An Appraisal of Critical Legal Challenges. *Space Policy*, 52, 101372.

Pandit, R. (2019). Eye on China, India Set to Kickstart 1st Space War Drill. *Times of India*, 24 July. Available at: https://timesofindia.indiatimes.com/india/eye-on-china-india-set-to-kickstart-1st-space-war-drill/articleshowprint/70354760.cms.

Paracha, S. (2013). Military Dimensions of the Indian Space Program. *Astropolitics*, 11(3), 156–186.

Pollpeter, K., Chase, M. and Heginbotham, E. (2017). *The Creation of the PLA Strategic Support Force and its Implications for Chinese Military Space Operations*. RAND Corporation. Available at: https://www.rand.org/pubs/research_reports/RR2058.html.

Pollpeter, K., Ditter, T., Miller, A. and Waidelich. B. (2020). *China's Space Narrative*. Montgomery, AL: China Aerospace Studies Institute, Air University. Available at: https://www.airuniversity.af.edu/CASI/Display/Article/2369900/chinas-space-narrative/.

Press Trust of India. (2019a). *Government Finalises Broad Contours of Defence Space Agency*, 11 June. Available at: https://economictimes.indiatimes.com/news/defence/government-finalises-broad-contours-of-defence-space-agency/articleshow/69745921.cms.

Press Trust of India. (2019b). *Opposition Parties Accuse PM Modi of 'Playing Politics' Over ASAT Missile Test*, 27 March. Available at: https://timesofindia.indiatimes.com/india/opposition-parties-accuse-pm-modi-of-playing-politics-over-asat-missile-test/articleshow/68599497.cms.

Raghuvanshi, V. (2019). India to Launch a Defense-Based Space Research Agency. *Defense News*, 12 June. Available at: https://www.defensenews.com/space/2019/06/12/india-to-launch-a-defense-based-space-research-agency/.

Raj, G. (2000). *Reach for the Stars: The Evolution of India's Rocket Programme*. New Delhi: Viking.

Rajagopalan, R.P. (2011). India's Changing Policy on Space Militarization: The Impact of China's ASAT Test. *India Review*, 10(4), 354–378.

Rajagopalan, R.P. (2020). India's Space Strategy: Geopolitics is the Driver. *ISPI Online*, 11 December. Available at: https://www.ispionline.it/en/pubblicazione/indias-space-strategy-geopolitics-driver-28607

Reddy, S. V. (2008). India's Forays into Space: Evolution of Its Space Programme. *International Studies*, 45(3), 215–245.

Reuters. (2020). *India Says to Sign Military Agreement with U.S. on Sharing of Satellite Data*, 26 October. Available at: https://www.reuters.com/article/us-india-usa-defence-idUSKBN27B1QY.

Samson, V. (2010). India's Missile Defense/Anti-Satellite Nexus'. *The Space Review*, 10 May. Available at: http://www.thespacereview.com/article/1621/1.

Sankar, U. (2007). *The Economics of India's Space Programme: An Explanatory Analysis*. New Delhi: Oxford University Press.

Sarabhai, V. (1968). Speech Given at the Dedication of the Equatorial Rocket Launching Station, *Thumba, Ind*ia, 2 February. In: P. V. Manoranjan Rao and P. Radhakrishnan (2012), *A Brief History of Rocketry in ISRO* (pp. 36–41). Hyderabad: Universities Press.

Sarabhai, V. (1974). *Science Policy and National Development*. Ed. K. Chowdhry. New Delhi: Macmillan.

Schaffer, T. C. and Schaffer, H. B. (2016). *India at the Global High Table: The Quest for Regional Primacy and Strategic Autonomy*. Washington, DC: Brookings Institution Press.

Shambaugh, D. (2004). *Modernizing China's Military: Progress, Problems, and Prospects*. Berkeley: University of California Press.

Sheehan, M. (2007). *The International Politics of Space*. Abingdon: Routledge.

Siddiqi. A.A. (2015). Making Space for the Nation: Satellite Television, Indian Scientific Elites, and the Cold War. *Comparative Studies of South Asia, Africa and the Middle East*, 35(1), 35–49.

Sil, R. and Katzenstein, P.J. (2010). *Beyond Paradigms: Analytic Eclecticism in the Study of World Politics*. Basingstoke: Palgrave Macmillan.

Singh, G. (2017). *The Indian Space Programme: India's Incredible Journey from the Third World Towards the First*. London: Astrotalkuk Publications.

Si-soo, P. (2022). Quad Nations Unveil Satellite-Based Maritime Monitoring Initiative. *SpaceNews*, 24 May. Available at: https://spacenews.com/quad-nations-unveil-satellite-based-maritime-monitoring-initiative/.

State Council of the People's Republic of China. (2015). *China's Military Strategy*. Beijing: The State Council Information Office of the People's Republic of China. Available at: http://english.www.gov.cn/archive/white_paper/2015/05/27/content_281475115610833.htm.

State Council of the People's Republic of China. (2019). *China's National Defense in the New Era*. Beijing: Foreign Languages Press Co. Ltd. Available at: http://www.china-un.ch/eng/dbtyw/cjjk_1/cjjzzdh/t1683060.htm.

Stroikos, D. (2018). Engineering World Society? Scientists, Internationalism, and the Advent of the Space Age. *International Politics*, 55(1), 73–90.

Stroikos, D. (2020). China, India, and the Social Construction of Technology in International Society: The English School Meets Science and Technology Studies. *Review of International Studies*, 46(5), 713–731.

Stroikos, D (2022a). Power Transition, Rising China, and the Regime for Outer Space in a US-Hegemonic Space Order. In: T. B. Knudsen and C. Navari (Eds.), *Power Transition in the Anarchical Society: Rising Powers, Institutional Change and the New World Order* (pp. 329–352). Cham: Palgrave Macmillan.

Stroikos, D. (2022b). International Relations and Outer Space. *Oxford Research Encyclopedia of International Studies*. Available at: https://oxfordre.com/internationalstudies/view/10.1093/acrefore/9780190846626.001.0001/acrefore-9780190846626-e-699.

Stroikos, D. (forthcoming). Still Lost in Space? Understanding China and India's Anti-Satellite Tests Through an Eclectic Approach. *Astropolitics*.

Suzuki, S. (2008). Seeking 'Legitimate' Great Power Status in Post-Cold War International Society: China's and Japan's Participation in UNPKO. *International Relations*, 22(1), 45–63.

Tellis, A.J. (2007). China's Military Space Strategy. *Survival*, 49(3), 41–72.

Tripathi, R. (2019). DRDO Ex-chief Saraswat Rules Out Political Motive. *The Indian Express*, 28 March. Available at: https://indianexpress.com/article/india/drdo-ex-chief-saraswat-rules-out-political-motive-5646284/.

Weeden, B. (2014). *Through a Glass, Darkly: Chinese, American, and Russian Antisatellite Testing in Space*. Washington, DC: Secure World Foundation. Available at: https://swfound.org/media/167224/through_a_glass_darkly_march2014.pdf.

Weeden, B. and Samson, V. (2021). *Global Counterspace Capabilities: An Open Source Assessment*. Washington, DC: Secure World Foundation. Available at:https://swfound.org/media/207162/swf_global_counterspace_capabilities_2021.pdf.

Wortzel, L.M. (2013). *The Dragon Extends its Reach: Chinese Military Goes Global*. Washington, DC: Potomac Books.

Wu, X. (2022). *China's Ambition in Space: Programs, Policy and Law*. The Hague: Eleven.

Wu, X. and Long, J. (2022). Assessing the Particularity and Potentiality of Civil–Military Integration Strategy for Space Activities in China. *Space Policy*, 62, 101514.

Xinhua. (2017). Backgrounder: Xi Jinping's vision for China's space development. *Xinhua*, 24 April. Available at: http://www.xinhuanet.com/english/2017-04/24/c_136232642.htm.

Xinhua. (2022). Xi Encourages Youth to Help Boost China's Aerospace Sci-tech Self-Reliance. *Xinhua*, 3 May. Available at: https://www.chinadaily.com.cn/a/202205/03/WS6270b55ca310fd2b29e5a783.html.

Zhang, H. (2005). Action/Reaction: U.S. Space Weaponization and China. *Arms Control Today*, 35(10), 6–11.

Zhang, B. (2011). The Security Dilemma in the U.S.-China Military Space Relationship: The Prospects for Arms Control. *Asian Survey*, 51(2), 311–332.

Zhao, L. (2014). Xi Calls for Joining Space and Air Roles. *China Daily*, 15 April. Available at: https://www.chinadaily.com.cn/china/2014-04/15/content_17433504.htm.

11 Military Strategy in Outer Space

A Call to Arms Control

Jessica West

Introduction

Although the "peaceful use of outer space" is a well-worn mantra of international governance ("COPUOS" 2021), a shift to warfighting is well underway among national militaries. Public statements are rife with rhetorical drumbeats of space as a "warfighting domain" (Smith 2017). From the United States to France, India, and Japan, militaries are creating new doctrines, strategies, and operational units to respond to the growing array of threats to space-based systems, both real and perceived. The stakes are high. The use of outer space systems and the data that they provide are essential to almost every function of modern militaries. But such use of space is also deeply integrated into the everyday lives of civilians all around the world. A war in outer space would have grave military, environmental, and humanitarian consequences (International Committee of the Red Cross 2021). Despite the risks, few mechanisms exist to prevent conflict escalation in space, and almost none that guard against harmful destruction in this sensitive and essential environment (United Nations 1967).

Focused on the United States, this chapter reviews the latest military doctrine and strategies that underpin warfighting in outer space. Aimed at maintaining strategic stability, the strategy flows through various layers of deterrence beginning with traditional military superiority, followed by capabilities for "deterrence through denial," such as detection, "active defenses" and resilience, and coming to rest on a commitment to "winning war" in space (United States Space Force 2020). But absent a strategy of political engagement and arms control, this series of one-sided technical fixes for warfighting resemble triage more than stability. It is likely to end in failure. Strategic stability and the prevention of conflict in outer space will not be met through technical means alone. These are political objectives that require political solutions. None are in sight. Although a nod to politics is evident in the inclusion of efforts to shape norms of behavior in outer space, here too the strategy remains one-sided, rooted in competition rather than cooperation. Missing from the current formula is a commitment to arms control.

This chapter is a critique of space power and defense strategies predicated on deterrence. Drawing on a broad interpretation of arms control as a form of political engagement among potential adversaries pioneered by Thomas Schelling

DOI: 10.4324/9781003230670-15

(Schelling 1961), I argue that it is a compliment to deterrence, and a necessary ingredient to achieve strategic stability and the prevention of conflict in outer space.

Revisiting Arms Control in Space

The track record of arms control in outer space is poor. Forty years after the first United Nations (UN) resolutions to prevent an arms race in outer space (PAROS), the international community is still trying to agree on how to do this (West 2021a). But long before the UN took up the mantra of PAROS, space was a focus of strategic and bilateral arms control initiatives. Some succeeded. Following disastrous tests of nuclear weapons that inflicted significant harm on the surrounding space environment and satellites, the Partial Test Ban Treaty banning nuclear explosions under water, in the atmosphere, and in outer space was first ratified by the United States, Soviet Union, and United Kingdom in 1963. The value of arms control in space for strategic stability is clearly demonstrated through bilateral talks, such as the Strategic Arms Limitation Talks (SALT), which produced both the Anti-Ballistic Missile (ABM) Treaty and the SALT I agreements, each of which prohibited interference with "national technical means of verification," (United States and Union of Soviet State Republics 1972a, Article V), widely understood to mean military reconnaissance satellites, as well as strategic communications capabilities in outer space (Koplow 2014). Additionally, the ABM Treaty limited the use of ABM systems to defend against ballistic missiles and restricting their placement in outer space (United States and Union of Soviet State Republics 1972b, Article V). But additional efforts to extend military restrictions in space – including a series of bilateral discussions in the 1970s and 1980s to ban anti-satellite (ASAT) weapons – failed. And in 2002, the United States formally withdrew from the ABM Treaty.

The U.S. national policy is not against arms control in space, in theory. In 1984, the U.S. Congress introduced "A joint resolution calling upon the President to seek a mutual and verifiable ban on weapons in space and on weapons designed to attack objects in space" (United States Congress 1984). The 2010 U.S. National Space Policy stipulated that it "will consider proposals and concepts for arms control measures if they are equitable, effectively verifiable, and enhance the national security of the United States and its allies" (Barak Obama 2010, 7).

Yet U.S. domestic space policy has been deeply resistant to the pursuit of arms control in outer space in practice. Military space strategy has been strongly influenced by theories of sea power (Bowen 2020; Leissle 2016). Much like space today, naval power has long been viewed as an enabling power for the projection of military force, underpinned by concepts of freedom, command, and control (Gray 1994). Naval forces have notoriously remained outside of strategic arms control initiatives. While obstacles to formal arms control agreements in outer space are well known and largely technical, with a focus on definitions and verification, these can and have been overcome before (Standfield 2021). Ultimately, the core objection to the pursuit of arms control is political and ideological.

Key theorists of space power are deeply resistant to arms control, not only in space but also in principle. Pioneering theorists of space power include Colin Gray, who famously viewed arms control as a paradox or "house of cards," concluding that it is "either impossible or unimportant" (Gray 1992, 16–19). Despite the unique physical environment of outer space, he famously insisted that when it comes to strategy, there is nothing unique about space (Klein 2021). Likewise, Everett Dolman is well known for his work sketching the inevitability of weapons and warfare as a result of geopolitical competition in outer space; a line of argument that has been taken up by others (Dolman 2002; Pavelec 2012). Promoters of arms control in space have been derided as naïve, hypocritical, or unpatriotic ("Whither Arms Control in Outer Space? Space Threats, Space Hypocrisy, and the Hope of Space Norms" 2020).

Grey's assessment about the futility of arms control rests not on strategy but on politics: That the relationship is, at its core, hostile (Gray 1992). This conclusion speaks to a broader argument by Robert Jervis that arms control is a band-aid style that fails to take into account the political drivers of conflict (Jervis 1993). Others note that arms control too often fails to produce stability in a crisis (Brooks 2020). But I believe that these analysts, who confuse outcome with process, are not looking at arms control in the right way.

Although it has become synonymous with the formal, legally binding arms control agreements of the Cold War, arms control has a broader meaning in national security strategy. Thomas Schelling described it as "all the forms of military cooperation among potential enemies that may reduce the risk of war, its scope and violence if it occurs, or the costs of being prepared for it" (Schelling 1961, 723). In essence, arms control is a tool for governing relations between potential enemies based on mutual accommodation and restraint. Aimed at preventing and mitigating the consequences of violent conflict, arms control measures can take both a hardware approach by controlling or restricting the development, production, and use of certain types of weapons, or a behavioral approach that seeks to enhance the transparency of military activities in an effort to avoid miscommunication, misinterpretation, and the unintended escalation of conflict. Most agreements blend the two.

Going forward, arms control must be about function not form. According to Martha Finnemore and Duncan Hollis, who have written about international cyber norms, the process is the product (Finnemore and Hollis 2016). Arms control includes a host of *interactive processes* and mechanisms from political engagement, dialogue, and negotiation to implementation measures that include communication, consultation, and forums to review and discuss compliance. These processes – not the agreement – are what give ongoing value to arms control. In a world where formal arms control agreements are in decline, the value of political engagement and less formal means of restraint are more essential than ever (Tannenwald 2020).

There is no time to wait. Writing in 2006 on the need for cooperative threat reduction in space to mitigate heightening distrust and a turn to arms, Clay Moltz noted, "fortunately, serious threats to security in space do not yet exist" (Moltz 2006, 121). But the subsequent shift among militaries to consider space a warfighting domain means that this benign assessment is no longer the case.

The Shift to Warfighting in Outer Space

The creation in 2020 of the Space Force as the sixth branch of the U.S. military solidified the growing treatment of outer space as an operational domain of warfighting. This development followed the re-establishment in August 2019 of the U.S. Space Command as a geographic combatant command within the U.S. armed forces. A summary of U.S. defense space strategy describes space as a hostile environment marked by great power competition (U.S. Department of Defense 2020, 3, 7). But not all capabilities are defensive. The U.S. military intends to develop and field offensive capabilities that target an adversary's space and counterspace capabilities for use in fighting and winning wars in and through space (United States Space Force 2020, 36).

This focus on warfighting marks a decisive shift in the military treatment of outer space, although space and the objects that humans send into space have long been critical elements in war plans (Bowen 2020). From the development of the first space launch capabilities and the deployment of the first artificial satellite in outer space, major state militaries have included space in their strategies. And while many lament the loss of what they view as "sanctuary" in space (United States Space Force 2020, 7), others claim that such an idyllic past never existed (Dickey 2020). Still, until recently, the dominant approach to space as a battleground has been one of restraints (Moltz 2008).

What is new is the clear mobilization of force in outer space. The United States is not alone. Many states are in various stages of developing, testing, and fielding kinetic, directed-energy, and cyber weapons that will be able to target systems in outer space (Weeden and Samson 2023). Years of restraint in testing ASAT systems ended with China's infamous 2007 kinetic ASAT test. The United States conducted an ASAT demonstration – Operation Burnt Frost – in 2008. India followed suit in 2019 (Mission Shakti). In 2020, the United States and United Kingdom accused Russia of releasing a "projectile" from a satellite in orbit in what they called a weapons test in space (West 2020). The use of other tactics to temporarily deny the use of space systems and information, such as jamming, is rampant (Weeden and Samson 2021).

These physical demonstrations of force are integrated with warfighting strategies and organizational structures. Within the last decade, the militaries of Russia and China have re-organized to incorporate space into more traditional warfighting functions. Russia created its Aerospace Forces; China established the Strategic Support Force, which integrates space, cyberspace, and electronic warfighting components to function in what it describes as "informationized" warfare (State Council Information Office 2019, 6).

Such trends are accelerating. The United Kingdom, France, Italy, and Germany host military space commands. Japan's Air-Self-Defense Force contains the Space Operations Squadron. India has constructed a Defence Space Agency to take over the space-related operations of its three armed forces (Army, Navy, Air Force).

More and more, space is driving military cooperation between the United States and its allies. For example, the Five Eyes intelligence allies – Australia, Canada, New Zealand, the United Kingdom, and the United States – share space situational awareness (SSA) data and intelligence. (This alliance could soon expand to formally include Japan). Operational cooperation is also increasing. The annual Schriever Wargames training event includes participants from the Five Eyes alliance as well as France, Germany, and Japan (Everstine 2021). Such cooperation extends to the NATO – the key link between Europe and the United States. In 2019, NATO formally declared space an "operational domain"; emphasized the need to protect civilian and military assets in space; and adopted a new, classified Space Policy (NATO 2021). The approach newly situates space as a "priority, rather than an afterthought" (Stickings 2020). While NATO has resisted direct references to warfighting, and its Secretary-General Jens Stoltenberg has stated that NATO would not "weaponize" space (Banks 2019), in 2021, the Alliance issued a statement formally extending the collective defense provisions of Article 5 of its founding treaty to include "attacks to, from, or within space" on a "case-by-case basis" (Atlantic Council 2021).

Clearly, the United States and its European partners expect – and are preparing for – space to feature in future conflict. But in the absence of arms control measures, current military strategy is ill-equipped to succeed.

Strategic Stability and the Arms Control Gap in Outer Space

An overarching objective of strategic stability belies the rhetorical focus on warfighting in outer space. The 2020 summary report of the U.S. Space Defense Strategy asserts, "The DoD desires a secure, stable, and accessible space domain, whose use by the United States and our allies and partners is underpinned by comprehensive, sustained military strength" (U.S. Department of Defense 2020, 1). This desire is echoed in the guiding principles of space power (United States Space Force 2020).

Deterrence is at the heart of this strategy. The point of space power, according to the U.S. Department of Defense (DoD), is to "deter and defeat aggression and protect national interests in space" to "avoid the application of force" if possible (United States Space Force 2020, vi). To this end, the military has adopted a layered approach to deterrence. It begins traditionally with emphasis on maintaining military superiority in space. Military might is reinforced by a policy of extended nuclear deterrence, clarified in the 2018 Nuclear Posture Review which included attacks on strategic space assets among the "significant non-nuclear strategic attack" that might invite a nuclear response (United States Office of the Secretary of Defense 2018, 21). Russia has adopted a similar policy (Bugos 2020).

But unlike traditional nuclear deterrence, discouraging a broad range of activities in an entire domain of operation, such as outer space, is complex (Langeland and Grossman 2021). Not all nefarious behaviors are strategic threats; some – such as jamming – are minor misdemeanors. Military strength is thus complemented with various approaches of denying adversaries the benefits of interfering with U.S. space capabilities that include superior detection and threat characterization abilities, the means to pre-empt possible attacks, and the ability to withstand harmful actions through resilience. Ultimately, however, deterrence lands on a commitment to winning a war in space, albeit one that the country seeks to avoid. The *2020 Defense Space Strategy* frames this as an ability to "compete, deter, and win in a complex security environment characterized by great power competition" (U.S. Department of Defense 2020, 1). The latest national security guidance rephrases this as "deter," "defend," and "defeat" (Joe R. Biden 2021, 14). This is a one-sided approach to both deterrence and stability based almost exclusively on technical capabilities.

Arms control is not a part of this strategy; it should be. Arms control is essential to both strategic stability and deterrence, which at their core involve relations with other states. To succeed, it requires both a believed threat to do something under certain circumstances, and a promise *not* to do something under other conditions. It is a strategy of engagement that requires that each party trust the other(s) (Krepon 2020). In other words, deterrence requires the mutual self-interest that marks arms control bargains (Schelling 1961). Strength alone is insufficient.

Strategic stability is classically defined as an outcome of deterrence and, like deterrence, it is much more complex today (Trenin 2019). Narrowly, it involves a relationship where neither state has an incentive to strike first (Schelling 20), but broader interpretations include an absence of armed conflict or even harmonious relations among nuclear powers (Acton 2013). Given the larger range of actors and capabilities that define international security relations today, strategic stability must be conceptualized in these broader terms (Acton 2013). An ongoing process of arms control rooted in engagement and mutual restraint is essential.

The continued relevance of arms control measures for strategic stability can be seen in the current bilateral Strategic Stability Dialogue between the United States and Russia, which aims to "lay the groundwork for future arms control and risk reduction measures" to ensure predictability and reduce the risk of arms control and nuclear war (Office of the Spokesperson 2021). A similar process is needed for space. Not only are space-based military capabilities strategically critical to any future conflict but also the layers of deterrence mechanisms laid out to protect them will not work if arms control is not part of the defense equation in outer space. An assessment of these various layers indicates key limitations to each. In the absence of arms control, the net effect is one of falling dominoes cascading toward failure.

The Limits of Deterrence through Strength

Military superiority and the process of dissuading an action through promise of punishment are at the heart of deterrence in space. But strength alone is insufficient. On the one hand, responding with military force in outer space is undesirable, not least because of the negative environmental consequences that stand to inflict additional harm on all users of outer space. For this reason, the 2017 *National Security Strategy* stipulates that harmful actions "will be met with a deliberate response at a time, place, manner, and domain of our choosing" (Trump 2017, 31). As noted above, this includes potential nuclear escalation. Yet, there are reasons to doubt the ability of traditional deterrence to prevent conflict in space.

Advocates for deterrence argue that space is the same as any other domain (Pace 2012). If so, then we can use Earth's history of violent terrestrial conflict and nuclear near-misses as evidence that deterrence does not work that well. But deterrence might work even less well in space.

Cyberspace is an instructive example. The escalating number and intensity of cyberattacks on western states suggests that deterrence is not working (Soesanto and Smeets 2021). Indeed, the difficulty of applying deterrence to cyberspace is identified as a key hurdle for the new strategic stability talks that are intended to delve into this topic (Sanger 2021). Deterrence is traditionally aimed at specific threats. How does it work against an entire domain, when there are so many actors, not easily identified or detected, who persist in denying their actions (Rajagopalan 2019). Another question: At what level do activities cease to be seen as nuisances and become real threats? Grey-zone activities are difficult to deter without risking further escalation. The strategy is all the more dangerous because we are not even sure what such escalation might look like. U.S. defense policy clearly states that the DoD "has limited operational experience with conflict beginning in or extending into space" (U.S. Department of Defense 2020, 4). Differing conceptions of conflict escalation among adversaries further complicates matters (Stone 2020). There is a lot that we do not know about the dynamics of conflict and escalation in space (Harrison et al. 2017, 20). The danger is unacceptable when it is considered that deterrence in both the cyber and space spheres is now an extension of nuclear deterrence.

Indeed, this link to nuclear weapons is an implicit acknowledgement of the difficulty of achieving effective deterrence in space. For this reason, U.S. deterrence tactics in space incorporate not only the maintenance of superior military strength and threat of nuclear escalation but also superior detection capabilities (Gleason and Hays 2020). This is sometimes described as deterrence through attribution (Stone 2020). The idea is that detection denies adversaries the ability to act clandestinely. Such detection is based on the concept of space domain awareness, an evolution of SSA that is meant to prioritize real-time awareness and tracking of potential threats in a warfighting domain (Erwin 2019). If adversaries know that they are being observed, they are less likely to act aggressively, the thinking goes. And if they pursue aggressive behavior nonetheless, such detection means that they are less able to escape punishment.

However, detection is also central to another layer of deterrence strategy in space, namely deterrence through denial. The logic behind this approach is that an actor is less likely to pursue aggressive action if it will be denied the *benefits* of such action, or if it is unlikely to succeed (Gleason and Hays 2020). It is an implicit acknowledgement of the limits of traditional deterrence through strength and punishment.

The Limits of Deterrence through Denial

Underpinned by detection and attribution capabilities, U.S. military strategy seeks to deny an adversary the benefit of attack through two approaches: pre-emption and resilience. Both face critical limits.

The U.S. warfighting approach to military activities in outer space includes a focus on "offensive operations," which American space power doctrine describes as targeting "an adversary's space and counterspace capabilities, reducing the effectiveness and lethality of adversary forces across all domains" to gain the initiative and "neutralize adversary space missions before they can be employed against friendly forces" (United States Space Force 2020, 36).

This strategy of pre-emption is couched in terms of self-defense or "active defense," which extends beyond a traditional focus on protection to include preventive actions against a threat as it materializes (Harrison, Johnson, and Young 2021, 36). According to U.S. space power doctrine, these "actions to destroy, nullify, or reduce the effectiveness of threats holding friendly space capabilities at risk" include not only "reactive operations" in response to an initiated attack but also "proactive efforts to seize the initiative once an attack is imminent" (United States Space Force 2020, 36).

The United States is not alone in pursing such a strategy. France also refers to an operational framework for "self-defense" in space (The French Ministry for the Armed Forces 2019, 9). China has stated, "we will not attack unless we are attacked, but we will surely counter-attack if attacked," and emphasizes "containing and winning wars, and underscores the unity of strategic defense and offense at operational and tactical levels" (State Council Information Office 2019, 8).

Although advocates describe active defense as a "gray area" that can be interpreted as either defensive or offensive, "depending on one's perspective" (Harrison, Johnson, and Young 2021, 36), a prospective target clearly views it as offensive. What is true from all perspectives is that, like any strategy that contains a pre-emptive component to strike first at a perceived threat, it is inherently unstable, particularly when adopted by many competing actors.

A blurring of offensive and defensive capabilities and strategies means that it becomes unclear which actions constitute an imminent threat and other actors do not know when to respond pre-emptively in turn (Townsend 2020). The deployment of offensive capabilities – even if intended for defensive use – intensifies the security dilemma and provokes counter-responses by competitors, thus *increasing* strategic *instability* (Teeple 2020). Further complications arise from the potential dual-use of military and civilian capabilities, such as active debris removal, which some proponents of active defense advocate (Harrison, Johnson, and Young 2021). Confusion, caused by misinterpretation and miscommunication and amplified by the perceived advantage and necessity of striking first, is a key driver of conflict. Indeed, active defense in space risks recreating the type of security dilemma that nuclear arms control agreements seek to resolve. This strategy is directly in opposition to the objective of stability.

This is clearly dangerous since a pre-emptive strategy is unlikely to work in practice. Many of the proposed capabilities that are intended to intercept an attack in outer space rely on flawless intelligence, immediate reaction, perfect timing. In reality, the laws of physics that dictate and constrain operational responses in outer space will not produce such ideal circumstances. Although objects in orbit move quickly, maneuvering and outmaneuvering in space are slow and difficult (Reeseman and

Wilson 2020). Finally, let us not forget that the outcome of a pre-emptive attack will be crisis escalation and counter-attacks – in other words, war (Regehr 2020, 12). Put into practice, active defense brings about what it seeks to prevent.

Like the limits of deterrence through strength and punishment, this is acknowledged by the second strand of the approach to denial: resilience. Resilience has been a part of U.S. defense strategy in space for at least a decade (U.S. Department of Defense 2011). The United States seeks to achieve resilience, which it sees as the ability to "continue providing required capabilities in the face of system failures, environmental challenges, or adversary actions" (Air Force Space Command 2016), by focusing on six attributes of space systems: disaggregation, distribution, diversification, protection, proliferation, and deception. Together these amount to an ability to absorb and withstand attacks, which in theory denies an aggressor the benefit of an attack.

Resilience is essential to survive in outer space, a harsh operating environment rife with intentional, accidental, and natural threats. It is much less escalatory than pre-emption and active defense and can contribute to stability by reducing incentives for aggression (Townsend 2020). But it too has limits.

One problem is that resilience is expensive. The obvious way to withstand disruption to a critical system is through redundant or duplicated capability that can be used as backup. But space systems are complex, expensive, and not easily reproduced. In part, this can be achieved through cooperation and interoperability with allies. This is a key objective of NATO cooperation in outer space. But not all capabilities are easily replicated or replaced. Efforts to recreate a non-space option for military GPS – despite being much less effective – point to the continued vulnerability and limited resilience of key systems in space (Cardillo 2020).

Disaggregating capabilities of large space systems into constellations of smaller and cheaper parts is another approach to resilience. In this instance, the broader system might be able to withstand damage to one component, or more easily replace a damaged piece. It's a "strength in numbers" model (Linville and Bettinger 2020.). This idea is not new – consider the drive to constellations by commercial Internet providers such as Starlink – but putting theory into military practice has been slow. Moreover, both duplicating and disaggregating space systems add to growing safety and sustainability risks linked to congestion in space.

Finally, while resiliency cannot be repurposed as an offensive capability, it can be perceived by outsiders as a display of relative invulnerability and produce unwelcome reactions by them, which can contribute to overall strategic instability. Consider the pursuit of nuclear invulnerability through ballistic missile defense as a case in point. Again, in the absence of mutual agreements, there is a risk that such an approach may drive rather than mitigate instability.

Resilience also has physical limitations. The ability of any system to withstand or absorb harmful effects is determined by both the scope and duration of the threat and the physical qualities of the system itself. As well, damage is caused not only by the direct attack but also by the environmental damage caused by the attack, which can be equally devastating and difficult to mitigate. The cascading effects of space debris – known as the Kessler effect – illustrate this reality.

Rather than pursue a mutual approach to deterrence that engages with potential adversaries, the U.S. strategy emphasizes technical capabilities. It is a strategy that embraces failure, made explicit within U.S. defense strategy itself, which notes, "Should deterrence fail, military space forces are prepared to fight and win our Nation's wars, in space, from space, and to space" (United States Space Force 2020). But the ability to win a war in space is dubious. Arms control is necessary, not only to make deterrence work but also to provide a safety net when it does not.

The Limits of "Winning" a War in Space

There is little public documentation on how wars in outer space are to be won. I suspect it is because when it comes to warfighting in space, there is no such thing as winning. In this, space is not unique. Earlier in his career, Colin Grey asserted that victory in nuclear war was possible (Gray and Payne 1980). He later changed his mind (Cimbala 2021).

It is not clear what winning a war in space even means. Writing about the prospect of nuclear war, Ernie Regehr argues that such a war "would inflict such massive and unacceptable damage that only losers would emerge" (Regehr 2020, 14). Current law bars nuclear war in space under Article IV of the Outer Space Treaty (OST), but massive harm is almost certain even without the use of nuclear weapons, because of the fragility of the environment and the connectedness of actors and activities.

It can be said without hyperbole that what is at risk is everyone's ability to safely access and use outer space (West 2022). While military competition, arms races, and warfighting are not unique to outer space, the ways in which they unfold in this environment are. There is no separate zone for warfighting: The whole space domain is the battlefield.

The historical consequences of the use of weapons in this fragile environment have been deadly and indiscriminate. The first nuclear tests conducted in space by the United States in 1958 contaminated the Van Allen belts around Earth with additional radiation and disabled at least six satellites (Moltz 2008, 68). China's ASAT test in 2007 created the largest debris field to date, with all the pieces still up there in space. One of those pieces recently smashed into a Chinese military satellite (Wall 2021). The cascading effects of even low-debris events in space are becoming more consequential as space becomes more and more crowded (Boley and Byers 2021). Additional destructive events could make some orbits unusable (Kessler and Cour-Palais 1978).

The risks of warfighting in space extend far beyond space itself. The goal of exerting power in space is to have military effects on Earth (Bowen 2020). These effects could be devastating to civilians because the targets of warfighting activity would be both the many military space systems, which are dual-use and support civilian functions, as well as many commercial systems that serve military and non-military clients. If space becomes a warfighting domain, the International Committee of the Red Cross has declared that "the human cost of using weapons in outer space that could disrupt, damage, destroy or disable civilian or dual-use space objects is likely to be significant" (International Committee of the Red Cross 2021, 1). Civilian signals for GPS and Europe's Galileo that are used for navigating vehicles from ships to commercial aircraft and drones, are already a target of hostile forces, even during peacetime (Zhang and Zhu 2017).

The potential for nuclear catastrophe is terrifying. The command and control of nuclear weapons systems requires the use of space assets. These assets could be deliberately targeted or accidentally hit or disrupted during combat, setting off nuclear devastation. The entanglement of space deterrence with nuclear deterrence introduces further risk.

It seems obvious that preventing a war should take priority over winning one. In the event that violence does erupt, there should be measures in place that prevent the worst possible outcomes. And so we are back to the critical need for arms control in outer space, not only to make deterrence more effective but also to provide a safety net when it does not.

A Nod to Norms

The U.S. *2020 Defense Space Strategy Summary* acknowledges the limitations and instability produced by a technical focus on deterrence and winning wars in space when it includes an aim to develop "standards and norms of behavior in space favorable to U.S., allied, and partner interests" as part of an effort to shape the strategic environment in outer space (U.S. Department of Defense 2020, 8). But here too, the focus remains on deterrence, and a one-sided approach to rule setting.

Norms – also known as rules of behavior or rules of the road, rooted in shared values and societal expectations of appropriate conduct – are a key instrument in the arms control toolkit (West 2021b). While norms focus on behavior, rather than restrictions on weapons, they can serve as transparency and confidence-building measures (TCBMs), which support arms control agreements by building trust among the participants (Rajagopalan 2015). Without this trust, agreements will not be successfully implemented. The promotion of such TCBMs has long been linked to the prevention of an arms race in outer space (PAROS) at the UN First Committee on Disarmament and International Security. A focus on behavior, including efforts to prevent misperceptions and miscommunication, can also help to address what some see as a limitation of arms control – the ability to manage crisis escalation (Brooks 2020). Such was the objective of the international space "code of conduct" proposed by the European Union in 2014 (European Union 2014). It remains the goal of the latest UN initiative to develop norms of "responsible" behavior in space, led by the United Kingdom (United Nations General Assembly 2020). But these good effects will not be achieved if norms become another weapon in the competition for power in space, or "normfare" (Radu et al. 2021). Norms must be broadly accepted and applied.

Norms as employed in U.S. space defense strategy set out rules of the road that favor American interests, including an effort to "inform international and public audiences of growing adversarial threats in space" (U.S. Department of Defense 2020, 8). This emphasis on promoting norms that explicitly benefit the United

States treats norms as an extension of the battlefield, producing not only stability for the United States but also winning hearts and minds at home and abroad. This strategy could be seen in action when the United States and United Kingdom publicly accused Russia of testing a weapon in outer space (West 2020).

This approach to norms is deeply embedded in deterrence. The aim is to inflict a political – rather than physical – cost for aggressive action in space (Langeland and Grossman 2021). This is of course preferable to the use of force. But it falls short of the type of engagement and mutual security that underpins arms control. There is also concern that a focus on "responsible" behavior, such as avoiding the creation of long-lived debris, might be used to legitimize rather than restrain the use of non-kinetic weapons in space (Hitchens 2021; West 2021c). A one-sided approach to strategic stability through a technical focus on deterrence and a political focus on self-fulfilling norms is not sufficient to prevent warfighting in outer space.

A Call to Arms Control in Space

War increases instability. In a time of mounting tension and talk of warfighting, the pursuit of arms control becomes more relevant to military and national security interests that value stability and predictability. Just as we have been expanding definitions and strategies of deterrence, we can shift our approach to arms control to meet current needs (Williams 2019). Contemporary arms control debates spend too much energy on form. Does arms control restrict hardware or behavior? Is it legally binding or voluntary? Instead, the focus should be on function. We want a tool that will result in mutual understandings and reciprocal commitments that mitigate the drivers of conflict and the resorting to violence. This function is not found in a final document, but in the process that produces it.

This view of arms control is aligned with Schelling's broad conceptualization of arms control as a relationship between adversaries. When thought about in this light, arms control becomes not a patch on the wound created by violent conflict but a means of mitigating if not necessarily resolving it. Whether states pursue legal agreements, norms of behavior, common understandings, or transparency and conflict-prevention mechanisms, what is critical is the process of engagement and mutual effort to prevent a resort to violence in outer space. A recent report by the UN Secretary-General summarizing states' views on threats in outer space and opportunities for progress indicates that there is common ground on which to build such engagement (United Nations Secretary-General 2021).

Of course, a commitment to arms control by the United States and allies is no guarantee that others will engage or follow suit. That is alright. The pursuit of arms control is a compliment to, and not a replacement for, strategies of deterrence. Arms control can make strategic stability through deterrence and conflict prevention stronger, while leaving the basis for these strategies intact. But if deterrence fails, we are left at the mercy of our efforts to mitigate the drivers and outcomes of violence.

In the absence of arms control, military strategy in space will likely cascade toward conflict. The cost of war is simply too high to bear.

Bibliography

Acton, James M. 2013. "Reclaiming Strategic Stability." *Carnegie Endowment for International Peace*. February 2013. https://carnegieendowment.org/2013/02/05/reclaiming-strategic-stability-pub-51032.

Air Force Space Command. 2016. "Resiliency and Disaggregated Space Architectures." *White Paper*. Colorado Springs, CO: Peterson AFB; U.S. Air Force Space Command. https://www.afspc.af.mil/Portals/3/documents/AFD-130821-034.pdf?ver=2016-04-14-154819-347.

Atlantic Council. 2021. "Brussels Summit Communiqué Issued by the Heads of State and Government Participating in the Meeting of the North Atlantic Council in Brussels 14 June 2021." *North American Treaty Organization*. June 14, 2021. https://www.nato.int/cps/en/natohq/news_185000.htm.

Banks, Martin. 2019. "NATO Names Space as an 'Operational Domain,' But without Plans to Weaponize It." *Defense News*. November 20, 2019. https://www.defensenews.com/smr/nato-2020-defined/2019/11/20/nato-names-space-as-an-operational-domain-but-without-plans-to-weaponize-it/.

Boley, Aaron C., and Michael Byers. 2021. "Satellite Mega-Constellations Create Risks in Low Earth Orbit, the Atmosphere and on Earth." *Scientific Reports* 11 (1): 10642. https://doi.org/10.1038/s41598-021-89909-7.

Bowen, Bleddyn E. 2020. *War in Space: Strategy, Spacepower, Geopolitics*. Edinburgh: Edinburgh University Press.

Brooks, Linton F. 2020. "The End of Arms Control?" *Daedalus* 149 (2): 84–100. https://doi.org/10.1162/daed_a_01791.

Bugos, Shannon. 2020. "Russia Releases Nuclear Deterrence Policy." *Arms Control Association*. August 2020. https://www.armscontrol.org/act/2020-07/news/russia-releases-nuclear-deterrence-policy.

Cardillo, Robert. 2020. "We Need a Backup for GPS. Actually, We Need Several of Them." *Defense One*. December 1, 2020. https://www.defenseone.com/ideas/2020/12/we-need-backup-gps-actually-we-need-several-them/170391/.

Cimbala, Stephen J. 2021. "Colin Gray: An Enduring Legacy." *Comparative Strategy* 40 (2): 216–218. https://doi.org/10.1080/01495933.2021.1880847.

"COPUOS." 2021. *United Nations Office for Outer Space Affairs*. https://www.unoosa.org/oosa/en/ourwork/copuos/index.html.

Dickey, Robin. 2020. "The Rise and Fall of Space Sanctuary in U.S. Space Policy." Center for Space Policy and Strategy. The Aerospace Corporation. https://aerospace.org/sites/default/files/2020-09/Updated_Dickey_SpaceSanctuary_20200901_0.pdf.

Dolman, Everett C. 2002. *Astropolitik: Classical Geopolitics in the Space Age*. Cass Series-Strategy and History. London; Portland, OR: Frank Cass.

Erwin, Sandra. 2019. "Air Force: SSA Is No More; It's 'Space Domain Awareness'." *SpaceNews*. November 14, 2019. https://spacenews.com/air-force-ssa-is-no-more-its-space-domain-awareness/.

European Union. 2014. "EU Proposal for an International Space Code of Conduct, Draft." *Text. European External Action Service - European Commission*. March 31, 2014. https://eeas.europa.eu/topics/disarmament-non-proliferation-and-arms-export-control/14715/eu-proposal-international-space-code-conduct-draft_en.

Everstine, Brian W. 2021. "Space Force Increasing International Outreach as the Service Grows." *Air Force Magazine* (blog). March 19, 2021. https://www.airforcemag.com/space-force-increasing-international-outreach-as-the-service-grows/.

Finnemore, Martha, and Duncan B. Hollis. 2016. "Constructing Norms for Global Cybersecurity." *The American Journal of International Law* 110 (3): 425–479.

Gleason, Michael P., and Peter L. Hays. 2020. "Getting the Most Deterrent Value from U.S. Space Forces." Space Agenda 2021. Aerospace Corporation.

"Global Counterspace Capabilities." 2021. Washington, DC: Secure World Foundation. https://swfound.org/media/207162/swf_global_counterspace_capabilities_2021.pdf.

Gray, Colin S. 1992. *House of Cards: Why Arms Control Must Fail*. Cornell Studies in Security Affairs. Ithaca, NY: Cornell University Press.

—— 1994. "Sea Power: The Great Enabler." *Naval War College Review* 47 (1): 18–27.

——, and Keith Payne. 1980. "Victory Is Possible." *Foreign Policy* 39 (Summer 1980): 14–27. https://doi.org/10.2307/1148409.

Harrison, Todd, Zack Cooper, Kaitlyn Johnson, and Thomas G. Roberts. 2017. *Escalation and Deterrence in the Second Space Age*. Washington, DC ; Lanham, MD: Center For Strategic & International Studies ; Rowman & Littlefield.

Harrison, Todd, Kaitlyn Johnson, and Makena Young. 2021. "Defense Against the Dark Arts in Space: Protecting Space Systems from Counterspace Weapons." *Aerospace Security Project*. Center for Strategic and International Studies.

Hitchens, Theresa. 2021. "Pentagon Poised To Unveil, Demonstrate Classified Space Weapon - Breaking Defense Breaking Defense - Defense Industry News, Analysis and Commentary." *Breaking Defense*. August 20, 2021. https://breakingdefense.com/2021/08/pentagon-posed-to-unveil-classified-space-weapon/.

International Committee of the Red Cross. 2021. "The Potential Human Cost of the Use of Weapons in Outer Space and the Protection Afforded by International Humanitarian Law." Position paper submitted by the International Committee of the Red Cross to the Secretary-General of the United Nations on the issues outlined in General Assembly Resolution 75/36. https://front.un-arm.org/wp-content/uploads/2021/04/icrc-position-paper-unsg-on-resolution-A-75-36-final-eng.pdf.

Jervis, Robert. 1993. "Arms Control, Stability, and Causes of War." *Political Science Quarterly* 108 (2): 239–253. https://doi.org/10.2307/2152010.

Kessler, Donald J., and Burton G. Cour-Palais. 1978. "Collision Frequency of Artificial Satellites: The Creation of a Debris Belt." *Journal of Geophysical Research* 83 (A6): 2637. https://doi.org/10.1029/JA083iA06p02637.

Klein, John J. 2021. "Some Lessons on Spacepower from Colin Gray." *Naval War College Review* 74 (1): Article 7.

Koplow, David A. 2014. "An Inference about Interference: A Surprising Application of Existing International Law to Inhibit Anti-Satellite Weapons." *Georgetown University Law Center* 35: 737–827.

Krepon, Michael. 2020. "Heroes of Arms Control: Thomas Schelling." *Arms Control Wonk*. July 13, 2020. https://www.armscontrolwonk.com/archive/1209738/heroes-of-arms-control-thomas-schelling/.

Langeland, Krista, and Derek Grossman. 2021. "Tailoring Deterrence for China in Space." *RAND Corporation*. https://www.rand.org/pubs/research_reports/RRA943-1.html.

Leissle, Marius. 2016. "Social Constructivism and Integration: Re-Igniting European Identity - a Common Ground in Space?" In *Theorizing European Space Policy*, edited by Thomas C. Hoerber and Emmanuel Sigalas. Lanham, MD: Lexington Books.

Linville, Capt Dax, and Maj. Robert A. Bettinger. 2020. "An Argument against Satellite Resiliency: Simplicity in the Face of Modern Satellite Design," *Air and Space Power Journal* (Spring 2020): 43–53.

Moltz, James Clay. 2006. "Preventing Conflict in Space: Cooperative Engagement as a Possible U.S. Strategy." *Astropolitics* 4 (2): 121–129. https://doi.org/10.1080/14777620600910563.

—— 2008. *The Politics of Space Security: Strategic Restraint and the Pursuit of National Interests*. Stanford, CA: Stanford Security Studies.

NATO. n.d. "NATO's Approach to Space." *NATO*. July 8, 2021. http://www.nato.int/cps/en/natohq/topics_175419.htm.

Office of the Spokesperson. 2021. "Deputy Secretary Sherman to Lead U.S. Delegation in Strategic Stability Dialogue with Russian Federation." *United States Department of State*. July 23, 2021. https://www.state.gov/deputy-secretary-sherman-to-lead-u-s-delegation-in-strategic-stability-dialogue-with-russian-federation/.

Pace, Scott. 2012. "Strengthening Space Security: Advancing US Interests in Outer Space." *Harvard International Review* 33 (4): 54–60.

Pavelec, Sterling Michael. 2012. "The Inevitability of the Weaponization of Space: Technological Constructivism versus Determinism." *Astropolitics* 10 (1): 39–48. https://doi.org/10.1080/14777622.2012.647392.

President Joe R. Biden. 2021. "Interim National Security Strategic Guidance." *The White House*. https://www.whitehouse.gov/wp-content/uploads/2021/03/NSC-1v2.pdf.

President of the United States, Barak Obama. 2010. "National Space Policy of the United States of America." *The White House*. https://history.nasa.gov/national_space_policy_6-28-10.pdf.

Radu, Roxana, Matthias C. Kettemann, Trisha Meyer, and Jamal Shahin. 2021. "Normfare: Norm Entrepreneurship in Internet Governance." *Telecommunications Policy* 45 (6): 102148. https://doi.org/10.1016/j.telpol.2021.102148.

Rajagopalan, Rajeswari Pillai. 2015. "Relationship of Norms of Behaviour with TCBMs and Treaties, in Particular the Legal and Political Advantages and Disadvantages of the Different Types of Instruments." In *Regional Perspectives on Norms of Behaviour for Outer Space Activities*. Geneva: United Nations Institute for Disarmament Research (UNIDIR) https://www.unidir.org/files/medias/pdfs/rajeswari-pillai-rajagopalan-eng-0-502.pdf.

—— 2019. "Electronic and Cyber Warfare in Outer Space." 3. *Space Dossier*. UNIDIR.

Reeseman, Rebecca, and James R. Wilson. 2020. "The Physics of Space War: How Orbital Dynamics Constrain Space-to-Space Engagement." *Aerospace Corporation*. https://aerospace.org/sites/default/files/2020-10/Reesman_PhysicsWarSpace_20201001.pdf.

Regehr, Ernie. 2020. *Deterrence, Arms Control, and Cooperative Security: Selected Writings on Arctic Security*. Peterborough: North American and Arctic Defence and Security Network (NAADSN).

Sanger, David E. 2021. "Once, Superpower Summits Were About Nukes. Now, It's Cyberweapons." *The New York Times*, June 15, 2021, sec. World. https://www.nytimes.com/2021/06/15/world/europe/biden-putin-cyberweapons.html.

Schelling, Thomas C. 1961. "The Future of Arms Control." *Operations Research* 9 (5): 722–731.

—— 1960. *The Strategy of Conflict*. Nachdr. d. Ausg. 1980. Cambridge, MA.: Harvard University Press.

Smith, Marcia. 2017. "Top Air Force Officials: Space Now Is a Warfighting Domain." *Space-PolicyOnline.com*. May 17, 2017. https://spacepolicyonline.com/news/top-air-force-officials-space-now-is-a-warfighting-domain/.

Soesanto, Stefan, and Max Smeets. 2021. "Cyber Deterrence: The Past, Present, and Future." In *NL ARMS Netherlands Annual Review of Military Studies 2020: Deterrence in the 21st Century—Insights from Theory and Practice*, edited by Frans Osinga and Tim Sweijs, 385–400. NL ARMS. The Hague: T.M.C. Asser Press. https://doi.org/10.1007/978-94-6265-419-8_20.

Standfield, Emily. 2021. "Lessons from the Chemical Weapons Convention." *Ploughshares Monitor*. September 13, 2021. https://ploughshares.ca/2021/09/lessons-from-the-chemical-weapons-convention/.

State Council Information Office, The People's Republic of China. 2019. *China's National Defense in the New Era*. https://english.www.gov.cn/atts/stream/files/5d3943eec6d0a15c923d2036

Stickings, Alexandra. 2020. "Space as an Operational Domain: What Next for NATO?" *RUSI News Brief* 40 (9): https://static.rusi.org/stickings_web_0.pdf.

Stone, Christopher M. 2020. "Deterrence in Space: Requirements for Credibility | RealClearDefense." December 1, 2020. https://www.realcleardefense.com/articles/2020/12/01/deterrence_in_space_requirements_for_credibility_651410.html.

Tannenwald, Nina. 2020. "Life beyond Arms Control: Moving toward a Global Regime of Nuclear Restraint & Responsibility." *Daedalus* 149 (2): 205–221. https://doi.org/10.1162/daed_a_01798.

Teeple, Nancy. 2020. "Offensive Weapons and the Future of Nuclear Arms Control." *Canadian Journal of European and Russian Studies* 14 (1): 79–102.

The French Ministry for the Armed Forces. 2019. "Space Defence Strategy." Report of the *"Space" Working Group*.

Townsend, Brad. 2020. "Strategic Choice and the Orbital Security Dilemma." *Strategic Studies Quarterly* (Spring 2020): 64–90.

Trenin, Dmitri. 2019. "Strategic Stability in the Changing World." *Carnegie Moscow Center*. March 2019. https://carnegiemoscow.org/2019/03/21/strategic-stability-in-changing-world-pub-78650.

United Nations. 1967. Treaty on Principles Governing the Activities of States in the Exploration and Use of Outer Space, *Including* the Moon and Other Celestial Bodies

United States, and Donald Trump. 2017. "National Security Strategy of the United States of America." https://www.whitehouse.gov/wp-content/uploads/2017/12/NSS-Final-12-18-2017-0905.pdf.

United States Congress. 1984. "A Joint Resolution Calling Upon the President to Seek a Mutual and Verifiable Ban on Weapons in Space and on Weapons Designed to Attack Objects in Space." *H.J.Res.524*, https://www.congress.gov/bill/98th-congress/house-joint-resolution/524?s=1&r=40

United Nations General Assembly. 2020. "Reducing Space Threats through Norms, Rules and Principles of Responsible Behaviours : Resolution." United Nations. https://digitallibrary.un.org/record/3893851?ln=en.

United Nations Secretary-General. 2021. "Reducing Space Threats through Norms, Rules and Principles of Responsible Behaviours." United Nations Disarmament Yearbook. United Nations Office for Disarmament Affairs. https://doi.org/10.18356/9789210056700c009.

United States of America and the Union of Soviet Socialist Republics. 1972a. Interim Agreement between the United States of America and the Union of Soviet Socialist Republics on Certain Measures with Respect to the Limitation of Strategic Offensive Arms (START 1).

United States of America and the Union of Soviet Socialist Republics. 1972b. Treaty between the United States of America and the Union of Soviet Socialist Republics on the Limitation of Anti-Ballistic Missile Systems.

United States Office of the Secretary of Defense. 2018. "Nuclear Posture Review." https://media.defense.gov/2018/Feb/02/2001872886/-1/-1/1/2018-NUCLEAR-POSTURE-REVIEW-FINAL-REPORT.PDF.

United States Space Force. 2020. "Spacepower: Doctrine for Space Forces." Space Capstone Publication. https://www.spaceforce.mil/Portals/1/Space%20Capstone%20Publication_10%20Aug%202020.pdf.

U.S. Department of Defense. 2011. "National Security Space Strategy, Unclassified Summary." U.S. Department of Defense and Office of the Director of National Intelligence. https://www.dni.gov/files/documents/Newsroom/Reports%20and%20Pubs/2011_nationalsecurityspacestrategy.pdf.

—— 2020. "Defense Space Strategy Summary." U.S. Department of Defense.

Wall, Mike. 2021. "Space Collision: Chinese Satellite Got Whacked by Hunk of Russian Rocket in March." *Space.Com*. August 17, 2021. https://www.space.com/space-junk-collision-chinese-satellite-yunhai-1-02.

Weeden, Brian, and Victoria Samson. 2023. *Global Counterspace Capabilities*. Washington, DC: Secure World Foundation. https://swfound.org/media/207567/swf_global_counterspace_capabilities_2023_v2.pdf.

West, Jessica. 2020a. "Did Russia Test a Weapon in Space?" *Project Ploughshares* (blog). July 2020. https://ploughshares.ca/pl_publications/did-russia-test-a-weapon-in-space/.

—— 2020b. "Whither Arms Control in Outer Space? Space Threats, Space Hypocrisy, and the Hope of Space Norms." Presented at the Center For Strategic and International Studies Webinar on "Threats, Challenges, and Opportunities in Space," *Zoom*. April 6. https://2017-2021.state.gov/whither-arms-control-in-outer-space-space-threats-space-hypocrisy-and-the-hope-of-space-norms/index.html.

—— 2021a. "Outer Space." In *First Committee Briefing Book*, 32–33. New York: Reaching Critical Will. https://reachingcriticalwill.org/images/documents/Disarmament-fora/1com/1com21/briefingbook/FCBB-2021.pdf.

—— 2021b. "Norms, Space Security, and Arms Control." *Project Ploughshares* (blog). June 2021. https://ploughshares.ca/pl_publications/norms-space-security-and-arms-control/.

—— 2021c. "A Weapons Test Is The Wrong Way To Advance Norms On Responsible Behavior in Space." *Breaking Defense*. August 26, 2021. https://breakingdefense.com/2021/08/a-weapons-test-is-the-wrong-way-to-advance-norms-on-responsible-behavior-in-space/.

—— 2022. "From Peaceful Uses to Warfighting: The Dangers of the New Military Era in Space." In *Military Space Ethics*, edited by Rev. Nikki Coleman, 269–286. Howgate Publishing Limited.

Williams, Heather. 2019. "Asymmetric Arms Control and Strategic Stability: Scenarios for Limiting Hypersonic Glide Vehicles." *Journal of Strategic Studies* 42 (6): 789–813. https://doi.org/10.1080/01402390.2019.1627521.

Zhang, Tao, and Quanyan Zhu. 2017. "Strategic Defense Against Deceptive Civilian GPS Spoofing of Unmanned Aerial Vehicles." In *Decision and Game Theory for Security*, edited by Stefan Rass, Bo An, Christopher Kiekintveld, Fei Fang and Stefan Schauer, 10575: 213–233. Lecture Notes in Computer Science. Cham: Springer International Publishing. https://doi.org/10.1007/978-3-319-68711-7_12.

Conclusion

A European Third Way in Space

Thomas Hoerber

In this book, we have seen many examples of the militarisation of space. Lorna Ryan described the processes of space militarisation and how the concept of the battlefield can be an organising concept for an understanding of the militarisation of space. Frank Slijper showed us the hesitations of the EU, particularly its hesitation concerning the militarisation of space. He calls it 'defensive', i.e., non-aggressive, which is an interpretation of changing realities that have been described as a new European reality by many before (see Hoerber, Forganni, 2021). General Pascal Legai, a seasoned practitioner in the field, analysed overall militarisation trends in Europe, right up to the very institutions of the European Space Agency and the EU which have become drivers for this. Thomas Hoerber goes back in history to the exclusively peaceful roots of European space policy. He makes a connection between the founding European value of preserving peace, which leads to a call for making a constructive and peaceful proposal of the world of institutional emancipation of the European Space Agency to an International Space Agency. Clearly, this is the idealist hope of preserving peace via rules in space rather than having more military technology in space. Isabelle Sourbès-Verger looked at space debris and the fact that everything that moves at high speed can be a weapon. This is the case for most space debris which is largely uncontrolled and growing in number. A hardening-up space infrastructure and the addition of manoeuvring capabilities to most new space installations is her proposed consequence. Iraklis Oikonomou has shown that the European space industry has no qualms about benefitting from the militarisation of space. Could it be that an industrial profit rationale is leading the EU away from its peaceful roots or, conversely, could this even be necessary in order to protect its citizens? In some country case studies, we have seen different national space preferences. Antonio Calcara showed in the Italian case the increase of European defence spending through the backdoor of space. Do we not dare to use the term of European military engagement in space, or has space become only one point in a growing defence policy across Europe? The quick and united reaction to war in Ukraine might lead us to that conclusion. The example of Luxembourg, described by Helen Kavvadia, shows another aspect of space militarisation: dual-use technology, in which Luxembourg has developed an expertise and a regulating framework, as well as vibrant technological, financial, and academic

DOI: 10.4324/9781003230670-16

ecosystems, which have allowed the small Member State to punch above its weight in expanding its industry into space militarisation applications.

In the section on international space actors, Mariel Borowitz has led us to the elephant in the room, i.e., the United States and its longstanding military space programmes. The new term 'Space Domain Awareness' reflects the shift from space as a benign environment to space as a warfighting domain, corresponding to the creation of U.S. Space Command and U.S. Space Force. China and India as rising powers and their militarisation of space, in Dimitrios Stroikos' words, have further consolidated the notion that conflict in space is inevitable. His analysis of the consequences for Europe is particularly telling. Jessica West finishes by advocating new, more, and better rules for the militarisation of space. An arms control system and a code of conduct suggested in the early 2000s by the EU are examples (Mutscheler, Venet, 2012).

This leads us to the final question of whether the EU ought to participate in this militarisation process of space, which has been explained through many factettes in this book. In the last publication leading up to this book, the ESSCA – School of Management standing group for space policy research asked similar questions:

> Can the EU do without a security policy? Or was that a luxury only a rather minor international organisation of the late 20th century could afford? If the EU becomes a global player in the 21st century, what security arrangements does it need? Should the EU follow the logic of the nation states of militarisation or is a new way possible, even though we do not know what it looks like today?
>
> (Hoerber, Forganni, 2021: 183)

Hoerber and Forganni (2021) showed that security considerations are already part of European space policy and that security is becoming more and more important in European politics. This growing concern about European security is, however, not limited to space policy. Rather, space policy seems to reflect a more general trend, which Mario Telò already spoke about in 2006: the choice Europe faces is between remaining a civilian power or going down the road of militarisation, not only of space, but also of all its security-relevant policies.

Mario Telò's Europe: a Civilian Power[1]

In contrast to recent trends for more European militarisation, Telò emphasised Europe as a civilian power (see also Duchêne, 1972; Maull, 1990; Majone, 2009; Sweeney, 2022). Telò's definition is very perceptive here:

> A political entity can be termed a civilian power not only if it does not intend, but also if it is not able, for various historical or structural reasons, to become a classic politico-military power and pursues its international peaceful objectives using other methods.
>
> (Telò, 2006: 51)

Telò points out that contrary to the negative perspective of lacking military capabilities, Europe must be seen as a very effective and indeed powerful international actor, because it has shown that it is well able to turn to good account its well-versed abilities as a civilian power (Telò, 2006: 57, see also Hoerber, Bohas, Valdemarin, 2023). Again, in contrast to the US perspective of using its military as a direct power tool, the EU has agreed to increased military capabilities under the Petersberg tasks only as a means of making its civilian engagement more credible; that is, military intervention as a possibility of last resort, the existence of which might make opponents more susceptible to preceding peaceful exercise of influence (Telò, 2006: 75).

Telò also stresses alternative and more innovative ways of future development and cooperation than the military, for example, space exploration as in the Ariane and the Galileo projects. In its very European way, the latter is not intended primarily for military use, such as the US GPS system, but primarily for civilian use, importantly independent of US influence (Telò, 2006: 54, 176). So, the question may be asked here of whether confrontation is the most effective way of resolving conflicts on the international stage, let alone whether it is the best way, if the political objective is peace. Telò agreed with Jean Monnet that an equal partnership between the EU and the US may be a much more constructive way forward (Telò, 2006: 75).

Telò sees the EU developing as a serious alternative to the American Empire. But European dynamism for integration is indispensable if, on the one hand, Europe does not want to be colonised by outside powers, for example, America or Asia or, on the other hand, disintegrate from within because of waning legitimacy among its peoples (Telò, 2006: 161). Telò sees the European social model as a key component in this European emancipation process both from within and in the world (Telò, 2006: 154) – the middle way, with a social conscience and a responsible attitude towards economic growth (Telò, 2006: Chapter 3, see Sustainabilism and the European Environmental Conscience in Hoerber, Weber, 2022). Telò shows the mixed government of the EU to be a working compromise which is still sufficiently flexible to adapt to the considerable changes which a consolidation of the European way will necessitate (Telò, 2006: 266).

Therefore, there is an alternative to Europe becoming another military power on this planet. Telò says clearly that the EU should not follow others down that road (Telò, 2006: 54, 145). In addition to finding the soul of Europe, the EU should rather use its experience to influence more troubled international organisations to develop into an effective world-governance system – the UN being the prime example (Telò, 2006: 71). This book has shown that space may be the field in which Europe may want to take another route than looking at each other down the barrel of a gun.

Outlook

In previous publications, the ESSCA standing group for space policy research has shown that because of the civilian foundations of European integration, which Telò (2006) thought should be preserved, European space policy has not in fact undergone a process of militarisation (Hoerber, Forganni, 2021). The 'Weaponisation' (see Mutschler, Venet, 2012: 119) of space at European level has taken place. It is true that some Member States have taken initiatives in that direction (see Antoni, Giannopapa, Schrogl, 2021; Legai, 2021). This may well be because of the conceived security threat in space, as pointed out by Lorna Ryan (2021). Similar to the arguments put forward in the present book, Oikonomou doubts whether there is any real danger. From his point of view, the build-up of a threat scenario in space serves both the military-industrial complex and the European Commission, in their search for power and money (Oikonomou, 2021). What we can clearly see in these contradicting arguments is that a struggle is underway about what security and related concepts, such as militarisation and weaponisation, mean and particularly what they mean in a European discourse.

The world has changed since the last book of 2021. The Russia-Ukraine war has even had an impact on peaceful space in the latest Russian ASAT test on 15 November 2021 (Foust, 2021).The International Space Station still works with Russian cosmonauts working closely together with astronauts from all participating nations. On the contrary, one could pose the question of effectiveness of the militarisation of space. Has it and will it produce more security, or rather an arms race and greater tension? At the current point, we do not have an answer of whether realists have got it right (Waltz, 2000) or whether there remains hope in the ideal that reason and mutual understanding could produce a peaceful world (Keohane, 1995). What is important for Europe is the question whether the security rationale leading powerful nations on this planet should also apply to the EU or whether the EU is *sue generis*, something different and can lead humankind to new horizons.

In this context, two important questions have been asked in recent publications: firstly, if the security aspects become more and more important in European space policy, if the realist arguments win the day, if the EU follows the same road as other major nation states on this planet, are we not creating an even bigger monster in the EU by diminishing the role of smaller European Nation States in their traditional function of national defence? Secondly, can we accept the consequences, if our trust in others is proven wrong, and if the liberal promise of moderating institutions fails? If, as a consequence, there were armed conflict? (Hoerber, Forganni, 2021: 185).

The conclusion for the EU seems to be that security policy is not an end in itself:

> The EU will never be comparable, militarily, with the US, China, or other major nation states. However, one should also not neglect security policies in the EU. This may be the European compromise which we find currently in EU policies relating to dual-use of space assets and the hardening up of its space infrastructure, for example.
>
> (Hoerber, Forganni, 2021: 185)

Going back to the original purpose of European integration:

> (…) the EU is a civilisation project, not just one which needs to be able to destroy others if it came to a final confrontation. For European space policy

this means, dual-use is fine as an exclusively defensive option. Such may even become the purpose of the EU in the future, to guarantee the security of its citizens. Militarisation of space, i.e. for potentially aggressive purposes, is definitely not fine. (…) let us not forget the peaceful purpose and the civilisation aspiration of the EU, of which the European space policy ought to be an important element.

(Hoerber, Forganni, 2021: 185)

This book on the militarisation of (European) space policies has produced further and ample evidence that things are changing in the world. Space is only one example. Europe has its role to play. A civilian and peaceful power in space may very well change the equation of other realist powers. Let us hope that new horizons are more attractive than ancient struggles.

Note

1 The following page has been published in a similar version as a book review in: *Journal of European Integration,* Volume 29, Issue 4 (2007), p. 526.

Bibliography

Antoni, N., Giannopapa, C., Schrogl, K.-U. (2021), 'European space policy perspectives on space traffic management (STM)', in: Hoerber, T., Forganni, A. (eds.), *A Growing Security Discourse in European Space Policy*, Routledge, London, pp. 97–118.

Duchêne, F. (1972), 'Europe's role in a world of peace', in: Mayne, R. (ed.), *Europe Tomorrow: Sixteen Europeans Look Ahead*, Fontana, London, pp. 1–21.

Foust, J. (2021), 'Russia destroys satellite in ASAT test', *Spacenews*, November 15, 2021, available online: https://spacenews.com/russia-destroys-satellite-in-asat-test/ last consulted on 11 October, 2021.

Hoerber, T., Bohas, A., Valdemarin, S. (2023), *The EU in a globalised World*, Routledge, London.

Hoerber, T., Forganni, A. (eds.) (2021), *A Growing Security Discourse in European Space Policy*, Routledge, London.

Hoerber, T., Weber, G. (eds.) (2022), *The European Environmental Conscience in EU Politics – A Developing Ideology*, Routledge, London.

Hoerber, T., Weber, G., Cabras, I. (eds.) (2022), *The Routledge Handbook of European Integrations*, Routledge, London.

Keohane, R. O., Martin, L. L. (1995), 'The promise of institutionalist theory', *International Security*, 20(1) (Summer, 1995), pp. 39–51.

Legai, P. (2021), 'Space, an indispensable dimension to the credibility of EU external action', in: Hoerber, T., Forganni, A. (eds.), *A Growing Security Discourse in European Space Policy*, Routledge, London, pp. 153–170.

Majone, G. (2009), *Europe as a Would-Be World Power: The EU at Fifty*, CUP, Cambridge.

Maull, H. (1990), 'Germany and Japan: The new civilian powers', *Foreign Affairs*, 69(5), pp. 91–106.

Mutschler, M., Venet, C. (2012), 'The EU as an emerging actor in space security', *Space Policy*, 28, pp. 118–224.

Oikonomou, I. (2021), 'The strategic utilisation of the US in EU military space policy discourse', in: Hoerber T., Forganni A. (eds.), *A Growing Security Discourse in European Space Policy*, Routledge, London, pp. 40–55.

Ryan, L. (2021), 'Security and the discourse of risk in European space policy', in: Hoerber T., Forganni, A. (eds.), *A Growing Security Discourse in European Space Policy*, Routledge, London, pp. 75–96.

Sweeney, S. (2022), 'EU common security and defence policy', in: Hoerber, T., Weber, G., Cabras, I. (eds.), *The Routledge Handbook of European Integrations*, Routledge, London, pp. 427–455.

Telò, M. (2006), *Europe: A Civilian Power? European Union, Global Governance*, World Power, Palgrave Macmillan, Basingstoke.

Waltz, K. N. (2000), 'Structural realism after the cold war', *International Security*, 25(1) (Summer 2000), pp. 5–41.

Index

For Product Safety Concerns and Information please contact our EU
representative GPSR@taylorandfrancis.com
Taylor & Francis Verlag GmbH, Kaufingerstraße 24, 80331 München, Germany

www.ingramcontent.com/pod-product-compliance
Lightning Source LLC
Chambersburg PA
CBHW060256220326
41598CB00027B/4121